Tourism and the Night

Over recent decades, municipal authorities have promoted their cities as places boasting desirable night-time activities. Light festivals, museum lates, nightclubbing, and night markets extend the typical tourist experience into the night and have become a key part of the way some cities are branded. This anthology draws together research addressing the relationship between tourism and the night, facilitating a better understanding of nocturnal city destinations.

Tourism and the Night: Rethinking Nocturnal Destinations covers an array of different tourist activities taking place at night and a range of European cities. The challenges facing late-night workers, the relationship between tourists and residents, and the effects of local policies on the expansion of late-night entertainment are examined in the first part of the book. The latter part focuses on the significance of night-time events, addressing the rising popularity of light art festivals and established religious rituals. Ultimately, this ground-breaking collection of papers examines how the night has become an important setting for city tourism. This trend means there is a need to rethink the management of urban districts and destinations, but there are also important implications for our understanding and experiences of the urban night.

The chapters in this book were originally published as a special issue in the *Journal of Policy Research in Tourism, Leisure and Events.*

Andrew Smith is a Reader in the School of Architecture and Cities at the University of Westminster, UK. He is the author of *Events in the City: Using Public Spaces as Event Venues* (2016) and *Events and Urban Regeneration: The Strategic Use of Events to Revitalise Cities* (2012).

Adam Eldridge is a Senior Lecturer in Sociology at the University of Westminster, UK. He is the co-author of *Planning the Night Time City* (2009) and co-editor of *Exploring Nightlife: Space, Society and Governance* (2018). His research examines the night as a source of debates about public space, leisure, and identity.

Tourism and the Night

Rethinking Nocturnal Destinations

Edited by
Andrew Smith and Adam Eldridge

Routledge
Taylor & Francis Group
LONDON AND NEW YORK

First published 2021
by Routledge
2 Park Square, Milton Park, Abingdon, Oxon, OX14 4RN

and by Routledge
605 Third Avenue, New York, NY 10158

Routledge is an imprint of the Taylor & Francis Group, an informa business

British Library Cataloguing-in-Publication Data
A catalogue record for this book is available from the British Library

ISBN13: 978-0-367-69513-2 (hbk)
ISBN13: 978-0-367-69517-0 (pbk)
ISBN13: 978-1-003-14209-6 (ebk)

Typeset in Minion Pro
by codeMantra

Publisher's Note
The publisher accepts responsibility for any inconsistencies that may have arisen during the conversion of this book from journal articles to book chapters, namely the inclusion of journal terminology.

Disclaimer
Every effort has been made to contact copyright holders for their permission to reprint material in this book. The publishers would be grateful to hear from any copyright holder who is not here acknowledged and will undertake to rectify any errors or omissions in future editions of this book.

Contents

Citation Information vi
Notes on Contributors viii

Introduction: Tourism and the night: towards a broader understanding of nocturnal city destinations 1
Adam Eldridge and Andrew Smith

1 Decoding middle-class protest against low-cost nocturnal tourism in Madrid 10
Begoña Aramayona and Rubén García-Sánchez

2 Gentrification, tourism and the night-time economy in Budapest's district VII – the role of regulation in a post-socialist context 24
Gergely Olt, Melanie Kay Smith, Adrienne Csizmady and Ivett Sziva

3 Commuting and the urban night: nocturnal mobilities in tourism and hospitality work 37
Anna Plyushteva

4 Strangers in the night: nightlife studies and new urban tourism 52
Adam Eldridge

5 Nocturnal ritual activities in tourist development of pilgrimage cities 66
Marie-Hélène Chevrier

6 Residents versus visitors at light festivals in cities: the case of Barcelona 85
Raquel Camprubí and Lluís Coromina

7 Fairy tale tourism: the architectural projection mapping of magically real and irreal festival lightscapes 99
Jane Lovell and Howard Griffin

Index 115

Citation Information

The chapters in this book were originally published in the *Journal of Policy Research in Tourism, Leisure and Events*, volume 11, issue 3 (June 2019). When citing this material, please use the original page numbering for each article, as follows:

Introduction

Tourism and the night: towards a broader understanding of nocturnal city destinations
Adam Eldridge and Andrew Smith
Journal of Policy Research in Tourism, Leisure and Events, volume 11, issue 3 (June 2019) pp. 371–379

Chapter 1

Decoding middle-class protest against low-cost nocturnal tourism in Madrid
Begoña Aramayona and Rubén García-Sánchez
Journal of Policy Research in Tourism, Leisure and Events, volume 11, issue 3 (June 2019) pp. 380–393

Chapter 2

Gentrification, tourism and the night-time economy in Budapest's district VII – the role of regulation in a post-socialist context
Gergely Olt, Melanie Kay Smith, Adrienne Csizmady and Ivett Sziva
Journal of Policy Research in Tourism, Leisure and Events, volume 11, issue 3 (June 2019) pp. 394–406

Chapter 3

Commuting and the urban night: nocturnal mobilities in tourism and hospitality work
Anna Plyushteva
Journal of Policy Research in Tourism, Leisure and Events, volume 11, issue 3 (June 2019) pp. 407–421

Chapter 4

Strangers in the night: nightlife studies and new urban tourism
Adam Eldridge
Journal of Policy Research in Tourism, Leisure and Events, volume 11, issue 3 (June 2019) pp. 422–435

Chapter 5

Nocturnal ritual activities in tourist development of pilgrimage cities
Marie-Hélène Chevrier
Journal of Policy Research in Tourism, Leisure and Events, volume 11, issue 3 (June 2019) pp. 436–454

Chapter 6

Residents versus visitors at light festivals in cities: the case of Barcelona
Raquel Camprubí and Lluís Coromina
Journal of Policy Research in Tourism, Leisure and Events, volume 11, issue 3 (June 2019) pp. 455–468

Chapter 7

Fairy tale tourism: the architectural projection mapping of magically real and irreal festival lightscapes
Jane Lovell and Howard Griffin
Journal of Policy Research in Tourism, Leisure and Events, volume 11, issue 3 (June 2019) pp. 469–483

Contributors

BEGOÑA ARAMAYONA Autonomous University of Madrid (UAM), Spain.

RAQUEL CAMPRUBÍ University of Girona, Spain.

MARIE-HÉLÈNE CHEVRIER Université de Strasbourg, France.

LLUÍS COROMINA University of Girona, Spain.

ADRIENNE CSIZMADY Hungarian Academy of Sciences, Centre for Social Sciences, Institute for Sociology, Hungary.

ADAM ELDRIDGE University of Westminster, UK.

RUBÉN GARCÍA-SÁNCHEZ European University of Madrid, Spain.

HOWARD GRIFFIN University of Kent, UK.

JANE LOVELL Canterbury Christ Church University, UK.

GERGELY OLT Hungarian Academy of Sciences, Centre for Social Sciences, Institute for Sociology, Hungary.

ANNA PLYUSHTEVA Vrije Universiteit, Belgium.

ANDREW SMITH University of Westminster, UK.

MELANIE KAY SMITH Hungarian Academy of Sciences, Centre for Social Sciences, Institute for Sociology, Hungary.

IVETT SZIVA Corvinus University of Budapest, Hungary.

INTRODUCTION

Tourism and the night: towards a broader understanding of nocturnal city destinations

Adam Eldridge and Andrew Smith

This special edition emerged out of a two-day symposium titled 'Tourism and the Night' which was held at the University of Westminster, London in the summer of 2017. Organised by the editors of this edition, and our colleague Ilaria Pappalepore, the symposium brought together scholars working at the interface of tourism and nightlife studies. Delegates from Australia, North America and across Europe attended the event, confirming our sense that the night is becoming increasingly important to the experience, study, management and representation of tourism. The papers presented were heavily oriented towards urban, Western contexts and the challenges and benefits of urban tourism at night. The policy interventions instigated by various cities were discussed with papers addressing a wide range of debates, practices, policies and theories. Case studies from Barcelona, Mashhad, Sydney, Lisbon, Toronto and Berlin, amongst others, demonstrated the breadth of research in this field and how the night and tourism intersect. The papers presented in this issue represent only some of the case studies examined over those two days, but they draw out some of the dominant themes; pilgrimage, lighting and light festivals, the rights of hospitality employees, gentrification, urban commodification, relations between tourists and locals, and the shifting ways that both nightlife and tourism are conceived.

The symposium had several points of departure. In particular, *The Villes Europeennes Et Nuit Urbaine* colloquium, held at the University Paul Valery, Montpellier, in 2016 was influential. This earlier colloquium had featured several papers on the aesthetics and politics of light festivals and illumination events. Importantly, these events were not discussed as simply tourist events that just happened to occur at night, nor were they read as merely an extension of already existing daytime tourism patterns. Instead, and echoing the extensive work of Edensor (2012, 2015), both the night and tourism were positioned as central to the ways the events were planned, managed and experienced. The 'Touristified Everyday Life' conference in the spring of 2017 at the Georg Simmel Centre for Metropolitan Studies in Berlin featured papers along a similar vein. Attention was drawn to the ways that concerns about local-tourist relations, gentrification, city branding, and everyday life were becoming especially critical at night. Debates about urban tourism and related themes of regeneration and gentrification were explored extensively, but the important question of what is it about the night that matters here was central. The issues of overtourism and touristification in Amsterdam, Berlin and Barcelona are well known, but what happens when those issues play out in the night? Is it simply a case of tourism extending into the night, or does the night in itself represent something different about the ways cities are experienced by tourists, and planned or represented by policy makers?

These events and our 2017 symposium coincided with several important policy interventions. Across much of the west, night-mayors have been recently installed in major urban centres to better manage and promote night-time uses (Roberts, 2016). Paris, London, Berlin, Zurich and many other western cities now have night-mayors, commissioners or tsars who, echoing our earlier point, do not simply extend the already existing policies of the day into the night but rather raise questions about how the night poses new and unique (if not singular) challenges and opportunities. A series of reports have also been commissioned by these agents with tourism at night recognised as important in terms of how respective cities are branded and promoted (GLA, 2017; Stevens, 2016).

The point derived from these academic events and recent policy initiatives is that the night matters, and rather than thinking about tourism activity at night as a mere extension of the day, our symposium encouraged delegates to think about how existing discourses, policies and experiences of tourism are shaped in the night time. We wanted to explore how nightlife is developing in ways that echo some of the concerns tourism scholars have explored such as mobility, changing everyday lives, the use of public space and urban development. But we also wanted to remain mindful of how the discourses that have long structured our understanding of the night – such as risk, transgression and enchantment – intersect, overlap and challenge discourses of tourism. It is worth pausing on this point to explain the title of the symposium: Tourism and the Night. This was a deliberate choice and we avoided calling the symposium Tourism *at* Night for precisely the reasons discussed above. As the papers at the symposium and in this collection identify, some everyday tourism activities are indeed extending into the later hours, but the strategies we have for managing tourism might not always work after dark and the issues raised by tourism, especially in already dense neighbourhoods, might be not so much intensified, but actually quite different in the night. If the night poses unique challenges to the experience and management of tourism, how might tourism also be changing the ways we conceive of the night? For these reasons, Tourism and the Night sought to explore the disjunctures and points of interest between the two areas of inquiry.

City tourism and the urban night

The study of tourism is fairly well established in the United Kingdom and covers a broad and disparate body of scholarship. Scholars at the University of Westminster, including Robert Maitland (2010), Ilaria Pappalepore (2010), Johannes Novy (2018) and Andrew Smith (2016) have researched tourism in cities, focusing on the creative city, events and festivals, (new) urban tourism and 'off-the-beaten-track' tourism (Maitland & Newman, 2014). Themes such as regeneration, gentrification, the use and management of public urban spaces, and critical reflections on the tourist-resident or host–guest binary run through this work. Emily Falconer's work with Edensor (Edensor & Falconer, 2012) and her single-authored work on gender, tourism and emotions (Falconer, 2013) represent another strength and strand of the University's work on tourism. These bodies of research draw upon a range and constantly evolving set of ideas and research methods, with the work of urban scholars (Judd & Fainstein, 1999; Hannigan, 1998; Zukin, 1989), and critical work around rebranding, regeneration, the commercialisation of urban sites, and emotions informing the ways tourism is conceived (Colomb, 2013).

The study of the night is perhaps more fragmented, drawing on criminology (Hobbs, Hadfield, Lister, & Winlow, 2003), sociology, cultural studies, and geography (Nofre & Eldridge, 2018; Shaw, 2018), planning (Roberts & Eldridge, 2009), and the health sciences. If it were at all possible to speak of 'night studies' we could point to a wide range of topics from licensing (Hadfield, 2006) and alcohol and drug use (Measham, 2006) to youth culture, debates about morality (Hubbard & Colosi, 2015), critical work on theories of the night and its historical and cultural specificities (Koslofsky, 2011; Williams, 2008), and debates about changing patterns of time-use (Crary, 2013). Research into affects, ambiences and emotions associated with the night have also been examined (Brands, Schwanen, & Van Aalst, 2015; Hubbard, 2016) while another body of work explores lighting in relation to planning and urban design, as well as the effects of lighting on human and non-human animals (Meier, Hasenöhrl, Krause, & Pottharst, 2014; Sloane, Slater, & Entwistle, 2016). There is a further strand of night studies which emphasises issues around security and the politics of securitising cities at night (Brands, Schwanen, & Van Aalst, 2016), while another body of work focuses on the commercialisation and commodification of existing urban spaces after dark (Chatterton & Hollands, 2003; Malet Calvo, Nofre, & Geraldes, 2017).

Differences in subject, language, key texts and theories exist across and within both fields of inquiry, but where tourism and night studies overlap is the shared interest in how spaces are lived, experienced, managed, and, just as importantly, enjoyed. These and other points of connection became evident over the course of our symposium. A dedicated attempt to analyse tourism *and* the night indicates not only the ways these bodies of scholarship sometimes differ in approach and focus but also the points at which they meet and land on similar questions. The papers in this edition thus represent different strands of thinking about the night and tourism, and the myriad ways they generate new perspectives when considered together.

Night-time tourism

By way of introducing the papers, it is worth first providing an overview of the established and changing patterns of activities that typify night-time tourism. Just as studies of the urban night have overly fixated on traditional (hedonistic) nightlife, studies of tourism at night have also tended to focus on established night-time activities: drinking, eating, theatre, and other forms of entertainment. There has been some attention dedicated to the rising popularity of night markets, particularly in Asia (Hsieh & Chang, 2006; Lee, Chang, Hou, & Lin, 2008), and there is a substantial body of work on other activities associated with the night and tourism, such as alcotourism and sex-tourism (Bell, 2008; Pinke-Sziva, Smith, Olt, & Berezvai, 2019; Sanders, 2017; Thurnell-Read, 2012) . Since the 1990s there has also been academic recognition of tourist-oriented occasions such as hen and stag parties, music events, and other festivals typically occurring in the night (Smith, 2016). While their impact on cities has been documented, how cities are increasingly targeting tourists by emphasising their late-night offer has been less well examined. Wider, 'everyday' tourism activity at night has also been neglected in academic analyses.

As noted above, conceptually, we avoided thinking of the night as simply an otherwise empty time and space into which tourism has started to expand. There has nonetheless been an extension of traditional daytime tourism activities into the night. Museums, galleries, and other attractions (even zoos) are staying open later, in some instances all night. Later opening hours for cultural institutions have been introduced in many cities to satisfy multiple policy agendas. For example, hours are often extended to make cities feel safer and as ways of encouraging families into the city at night. Late opening facilitates tourism in several ways: it extends opportunities for conventional city tourists seeking to make the most of short stays. Extending opening beyond normal working hours also provides new ways for citizens to consume their own cities as tourists. In some parts of the world, late-night opening is an established practice, but it is spreading via the widespread introduction of initiatives like 'museum lates' (Evans, 2012).

Adopting late-night opening in other contexts is a temporal form of globalisation, one in which different schedules, routines and rhythms are diffused globally while overlapping and sometimes competing at the local level. The ultimate global metropolis is the 24-hour city – a status which renders variations in daylight hours across the world less relevant. New York has long sold itself as 'the city that never sleeps', and this is an attraction for night owls that enjoy the novelty of shopping or dining in the middle of the night. There is probably a less enthusiastic set of consumers who also use these services: 24-hour opening is welcomed by long haul tourists struggling to adapt to new time zones.

As several authors have pointed out, however, this extension of daytime activity into the night – and the extension of daytime tourism into the night – is perhaps contributing to an erosion of the traditional distinction between day and night. Indeed, authors such as Sandhu (2007) see the very idea of night time in global cities like London as a rather old fashioned concept, and one which is rapidly waning. It is notable that, in one of the few early studies into the emergence of the '24 hour city' Kreitzman's 'The 24 Hour Society' (1999) was one of the first accounts of this sense that the night and day were blurring and tourism played a key role here. Kreiztman regarded the tourist almost as the prototype of the new 24-hour citizen, one who expected to eat, shop, or be entertained at the hour of their own choosing, irrespective of the traditional Monday–Friday, 9.00 am–5.00 pm, of the industrial, modern city. This portentous conception of the new, 24-hour citizen reappeared some years later to justify 24-hour alcohol licensing in England and Wales. The former British Prime Minister Tony Blair spoke of confused tourists, fresh from a night at the theatre, unable to secure a drink due to Britain's formerly restrictive licensing hours (Roberts & Eldridge, 2009). If the figure of the tourist, free from work and quotidian concerns becomes the model 24-hour citizen, it is not only the tourist from elsewhere who benefits, of course, but also residents who might now increasingly consume the museums, theatres, bars and retail opportunities afforded to them in response to the tourist market.

In other instances, however, it is not so much the dissolving of the night and day that drives tourism, but that the night itself is the core attraction. Despite concerns about the gradual erosion of night time, the night still offers unique dimensions that can provide distinctive tourism experiences. Examples include star gazing and various forms of astro-tourism (Ingle, 2010) which are based on the visual appeal of the night sky. For obvious reasons, the stars are best observed from undeveloped zones – particularly 'dark sky reserves' – but several urban areas offer astro-tourism where star gazing can be combined with astronomical culture and history. Various cities also offer nocturnal wildlife tours and a range of dark tours (e.g. ghost tours) which operate at night. The night is intrinsically associated with sleep, but this dimension of the night has also now been commodified in recent years. To provide an added dimension, some attractions offer the opportunity for guests to sleep on the premises overnight: including museums, toy shops and burial sites. In most instances, these opportunities are provided during dedicated events, but some attractions have installed more permanent accommodation. For example, visitors can now stay overnight within London Zoo in themed Indian style accommodation next to the Lion enclosure. The Zoo also offers a Bed Bugs experience, where brave visitors can stay overnight in the Bug House to learn more about the creepy crawlies kept there. London Zoo's new focus on the night is representative of the wider 'nocturnalisation' (Koslofsky, 2011) of tourism in the U.K. capital. The Visit London website now includes 'the top ten things to do at night', a list that includes bat walks in the Royal Parks, star gazing at the Royal Observatory and sleepovers at various National Museums. Rather than merely comprising night time events, or representing an extension of tourism into the night, these attractions reconfigure the night as *event*. Added to this, we must, of course, recognise the various ways the night continues to be framed by narratives of transgression, freedom and the loosening of everyday restrictions. The bars and clubs of Berlin, Ibiza or London benefit greatly from discourses which reinstate the distinction between the day as work and the quotidian, and the night as risk, pleasure, and adventure.

A final dimension of tourism at night beyond the obvious involvement of tourists in the nightlife and entertainment sectors is the emergence of what Shaw (2018) has termed the spectacular night. This is encapsulated by the enduring appeal of 'bright lights, big city' (Shaw, 2018). The city as illuminated spectacle is something inextricably linked with New York, a destination that became popular with tourists because of its reputation as a nocturnal city of lights. New York helped to establish the commercial aesthetic of floodlighting and illuminated advertising (Gilbert & Hancock, 2006), features which continue to attract tourists to places like Hong Kong and Tokyo. Cities now spend a great deal of time and money on illumination; beautification now rivals security as the central objective of lighting strategies in some cities (Giordano, 2018). This means sightseeing is something no longer confined to daylight hours. Alongside permanent illuminations, there has also been a sharp rise in the number of cities hosting light festivals and other light-based events. Lantern festivals and candle-lit processions have long attracted tourists to cities, but more recently a whole series of new events have been established dedicated to light art and light projections which transform the built environment and blur the distinction between the real and virtual city. This trend is part of a wider spectacularisation of the city, and it provides further evidence of the way that city spaces are animated – even performed – via programmed events. As Colomb (2013) has argued, the contemporary city is not merely a stage for these types of events, the city is staged, with urban spaces not merely the setting but the core attraction. Attracting tourists is a major motivation for 'staging' cities, and light festivals have added appeal as tourism development initiatives as they help to encourage visits during darker time periods of the year/the day which tend to be less popular. This helps to explain the recent inauguration of new light festivals in a range of urban locations (Giordano & Ong, 2017).

What we come to here are different ways of understanding the intersection of tourism and the night. If the first suggests an expansion of otherwise established tourism offers into the dark, such as museums and galleries, the second continues to reiterate unique atmospheric and experiential aspects of the night and relies on darkness itself to function. Following this trajectory, the night isn't simply a backdrop, but is instead crucial to a range of desired effects and affects from lighting displays to club nights.

But if the challenges of managing high tourist numbers have been known for some time, these events and others render the night as a new ground for potential conflicts as well as opportunities. That is, while tourism has for some time now been recognised to offer both challenges and social and economic benefits, the effects of tourism at night have moved in two directions. As Novy (2018) has advised, the conflicts now strongly articulated with urban tourism are still relatively discreet and there is a danger of over-emphasising the problems caused by tourism. Much the same can be said of the night. Despite the continuing narrative of anti-social behaviour, binge drinking and violence that colours particularly the British conception of nightlife, these conflicts also remain fairly contained and are often representative of wider and ongoing tensions about the contested city – at both day and night (Eldridge and Roberts, 2008). As well as anti-social behaviour, where tourism and night intersect in this narrative is a concern about economic expansion, commodification, and commercialisation. These concerns are not exclusive to nightlife or tourism, but in heavily touristified cities at night, namely Berlin, Amsterdam or Barcelona, it is precisely that sense that local bars, restaurants, and nightlife areas are experiencing the very worst aspects of tourism after dark that motivates much anti-tourism feeling. Perhaps nothing more clearly documents the intersection of tourism and the night than the oft-cited conflict between residents attempting to sleep while neighbouring tourists, in rented apartments or downstairs bars, enjoy the city at night.

The first few papers presented here tend to steer more towards this negative conception of tourism and the night, but the positive aspects should not be discounted. On the one hand, the sheer number of tourists now circulating within particular sites and the resulting levels of concern that causes have been well documented and need recognising. This should not overshadow the opportunities that the night still offers, be that in the form of pleasure, adventure, socialising, and the experience of new atmospheres. Equally, we need to avoid homogenising the urban tourist, who, until now, has remained fairly absent from this introductory discussion. The urban tourist is too often regarded as a single entity, a privileged nomad who desires little more than being entertained. The types of tourist practices explored below hopefully illustrate a more nuanced take on this figure, one who comes with a range of desires; inclusion, prayer, fun, and enchantment. How these are enabled or constrained by their class, sexuality, race and gender further complicates the figure of the tourist at night. Night studies have been especially attuned to issues around inclusion and exclusion, especially in terms of race, class and gender (Stevenson, 2018; Talbot, 2016), corresponding with similar work which analyses tourism as an activity that can both reproduce and challenge privilege.

What becomes especially evident here is just as tourism is difficult to pin down and is deeply contextual, the night is also not a uniform thing. We might often think of the night as merely a natural phenomenon but, as Williams argues, 'Although night is part of the natural world, its social uses and meanings are not, arising as they do from social practices' (Williams, 2008, p. 516). These practices, informed by religion, culture, history, and the like shape the ways the night is lived by locals and tourists. What becomes imaginable, desired, promoted, hidden and lived varies substantially across different nightscapes. The papers below clearly point to the myriad ways these motivations become evident. The cases covered are almost exclusively European, but the ideas are relevant to wider urban contexts.

The opening paper, by Aramayona and García-Sánchez, takes us to Madrid where the neighbourhood of La Latina is caught between the desires of tourists and the demands of local residents. Aramayona and García-Sánchez focus on the issue of class which influences the way the local concerns about visitors are framed. This perspective means the paper goes beyond restating the obvious potential for conflict between locals and tourists. Tourist behaviour at night in La Latina does not correspond with the narrative of the neighbourhood promoted by local residents. Aramayona and García-Sánchez argue that the conflict recalls the regime of Spain's former dictator Franco in the ways it draws upon familiar tropes of disgust and immorality as a means of distinction.

The next paper by Olt, Smith, Csizmady and Sziva also challenges conventional wisdom by moving away from the dominant framework of neoliberalism in explaining the development of the night-time economy. Instead, the authors argue that the development, management and regulation of nightlife in the 'party quarter' of Budapest reflect the constraints and political struggles associated with post-socialist, neo-patrimonial governance.

By focusing heavily on tourist experiences, tourism studies tend to neglect the experiences of the workers involved in producing experiences. Plyushteva's paper looks at the difficulties experienced by late-night hospitality workers in Brussels and Sofia. There is growing awareness of the precarious nature of tourism employment and poor pay/conditions, but the simple question of how workers get home at night has not been properly examined and this paper advances our understanding of nocturnal commuting.

Eldridge's paper takes a more speculative turn and mediates on the figure of the tourist as a stranger in the night. He argues that a unique feature of tourism in the night is precisely our desire to not be strangers but instead to feel a sense of attachment and belonging. The use of local facilities by tourists has long been recognised by urban tourism scholars, but laundrettes and markets differ to neighbourhood bars. Motivated by a desire to reconfigure the us and them binary that too often frames understandings of night-time conflicts, the paper suggests that the ways that new urban tourism scholars have questioned the tourist-resident binary needs to inform more critical work of the night.

The final three papers focus on night events, ranging from more traditional candle-lit processions to the contemporary trend for festivals featuring light projections.

Chevrier's paper demonstrates the point made earlier about the importance of the night and darkness to some tourist experiences after dark. Focusing on sacred events in Lyon and Lourdes, this paper explores the experiences of pilgrims in sacred spaces, arguing that nocturnal events challenge the dominant conception of the night as a time for youth and inebriation. Current night-time policy seeks to expand entertainment into the night, rendering the night less important, but Chevrier demonstrates that darkness and the opportunities it offers for reflection are crucial to understanding the pilgrims of Lourdes and Lyon.

The paper by Camprubi and Coromina analyses the Llum BCN light festival which is staged annually in Barcelona. Light festivals are often hosted to attract tourists, but they also allow residents a chance to see their own city 'in a new light'. Accordingly, their study uses data from surveys to analyse differences between tourists and residents in terms of their motivation to attend, their attitudes towards different types of installations and the likelihood they will attend the festival again. The study reveals that light festivals can simultaneously satisfy both residents and tourists, albeit in different ways.

Lovell and Griffin's paper stems from an innovative collaboration between academics and practitioners; and between tourism/event studies and architectural analysis. Light shows featuring projections are often regarded as homogenous events when in fact there is a great deal of diversity in their form, content and impact. Lovell and Griffin contribute to our understanding of light projections by developing a three-way conceptualisation: architecturally passive, architecturally physically active and architecturally metaphysically active. These categories have implications for the ways light shows are consumed by tourists.

Together, the papers point to different ways the night and tourism intersect. The themes addressed in this issue point to differences in conceptual models for understanding the 'night' and 'tourism'. Brought together, opportunities for the expansion of interdisciplinary knowledge and ways of conceptualising tourism in the night have been explored. This special issue functions as a point of departure for further exploration of the night as a tourist time–space, and an expanding field of inquiry into how the night shapes tourism, and tourism shapes the night.

Disclosure statement

No potential conflict of interest was reported by the authors.

References

Bell, D. (2008). Destination drinking: Toward a research agenda on alcotourism. *Drugs: Education, Prevention and Policy*, *15*(3), 291–304.

Brands, J., Schwanen, T., & Van Aalst, I. (2015). Fear of crime and affective ambiguities in the night-time economy. *Urban Studies*, *52*(3), 439–455.

Brands, J., Schwanen, T., & Van Aalst, I. (2016). What are you looking at? Visitors' perspectives on CCTV in the night-time economy. *European Urban and Regional Studies*, *23*(1), 23–39.

Chatterton, P., & Hollands, R. (2003). *Urban nightscapes: Youth cultures, pleasure spaces and corporate power*. Abingdon: Routledge.

Colomb, C. (2013). *Staging the New Berlin: Place marketing and the politics of urban reinvention post-1989*. Abingdon: Routledge.

Crary, J. (2013). *24/7: Late capitalism and the ends of sleep*. London: Verso Books.

Edensor, T. (2012). Illuminated atmospheres: Anticipating and reproducing the flow of affective experience in Blackpool. *Environment and Planning D: Society and Space*, *30*(6), 1103–1122.

Edensor, T. (2015). The gloomy city: Rethinking the relationship between light and dark. *Urban Studies, 52*(3), 422–438.
Edensor, T., & Falconer, E. (2012). Sensuous geographies of tourism. *The Routledge Handbook of Tourism Geographies*, 74–81.
Eldridge, A., & Roberts, M. (2008). A comfortable night out? Alcohol, drunkenness and inclusive town centres. *Area, 40*(3), 365–374.
Evans, G. (2012). Hold back the night: Nuit Blanche and all-night events in capital cities. *Current Issues in Tourism, 15*(1–2), 35–49.
Falconer, E. (2013). Transformations of the backpacking food tourist: Emotions and conflicts. *Tourist Studies, 13*(1), 21–35.
Gilbert, D., & Hancock, C. (2006). New York City and the transatlantic imagination: French and English tourism and the spectacle of the modern metropolis, 1893–1939. *Journal of Urban History, 33*(1), 77–107.
Giordano, E. (2018). Outdoor lighting design as a tool for tourist development: The case of Valladolid. *European Planning Studies, 26*(1), 55–74.
Giordano, E., & Ong, C. E. (2017). Light festivals, policy mobilities and urban tourism. *Tourism Geographies, 19* (5), 699–716.
GLA. (2017). *From good night to great night: A vision for London as a 24-hour city*. London: Greater London Authority.
Hadfield, P. (2006). *Bar wars: Contesting the night in contemporary British cities*. Oxford: Oxford University Press.
Hannigan, J. (1998). *Fantasy city: Pleasure and profit in the postmodern metropolis*. Abingdon, Oxon: Routledge.
Hobbs, D., Hadfield, P., Lister, S., & Winlow, S. (2003). *Bouncers: Violence and governance in the night-time economy*. Oxford: Oxford University Press.
Hsieh, A. T., & Chang, J. (2006). Shopping and tourist night markets in Taiwan. *Tourism Management, 27*(1), 138–145.
Hubbard, P. (2016). The geographies of "going out": Emotion and embodiment in the evening economy. In J. Davidson, L. Bondi, & M. Smith (Eds.), *Emotional Geographies* (pp. 131–148). London: Routledge.
Hubbard, P., & Colosi, R. (2015). Respectability, morality and disgust in the night-time economy: Exploring reactions to 'lap dance' clubs in England and Wales. *The Sociological Review, 63*(4), 782–800.
Ingle, M. (2010). Making the most of 'nothing': Astro-tourism, the Sublime, and the Karoo as a 'space destination'. *Transformation: Critical Perspectives on Southern Africa, 74*(1), 87–111.
Judd, D., & Fainstein, S. (Eds.). (1999). *The tourist city*. New Haven, CT: Yale University Press.
Koslofsky, C. (2011). *Evening's empire: A history of the night in early modern Europe*. Cambridge, UK: Cambridge University Press.
Kreitzman, J. (1999). *The 24 hour Society*. London: Profile Books.
Lee, S. H., Chang, S. C., Hou, J. S., & Lin, C. H. (2008). Night market experience and image of temporary residents and foreign visitors. *International Journal of Culture, Tourism and Hospitality Research, 2*(3), 217–233.
Maitland, R. (2010). *Introduction: National capitals and city tourism*. City Tourism, National Capital Perspectives, Oxfordshire: CABI.
Maitland, R., & Newman, P. (Eds.). (2014). *World tourism cities: Developing tourism off the beaten track*. London: Routledge.
Malet Calvo, D., Nofre, J., & Geraldes, M. (2017). The Erasmus Corner: Place-making of a sanitised nightlife spot in the Bairro Alto (Lisbon, Portugal). *Leisure Studies, 36*(6), 778–792.
Measham, F. (2006). The new policy mix: Alcohol, harm minimisation, and determined drunkenness in contemporary society. *International Journal of Drug Policy, 17*(4), 258–268.
Meier, J., Hasenöhrl, U., Krause, K., & Pottharst, M. (Eds.). (2014). *Urban lighting, light pollution and society*. Abingdon, Oxon: Routledge.
Nofre, J., & Eldridge, A. (Eds.). (2018). *Exploring nightlife: Space, society and governance*. London: Rowman & Littlefield International.
Novy, J. (2018). Urban tourism as a bone of contention: Four explanatory hypotheses and a caveat. *International Journal of Tourism Cities, 5*(1), 63–74.
Pappalepore, I. (2010). *Tourism and the development of creative urban areas: evidence from four non-central areas in London* (Doctoral dissertation). London: University of Westminster.
Pinke-Sziva, I., Smith, M., Olt, G., & Berezvai, Z. (2019). Overtourism and the night-time economy: A case study of Budapest. *International Journal of Tourism Cities, 5*(1), 1–16.
Roberts, M. (2016). What a 'Night Czar' can do to help nightlife survive. *The Conversation*. Retrieved from https://theconversation.com/what-a-night-czar-can-do-to-help-nightlife-survive-67253
Roberts, M., & Eldridge, A. (2009). *Planning the night-time city*. Abingdon: Routledge.
Sanders, M. (2017). *Women and sex tourism Landscapes*. London: Routledge.

Sandhu, S. (2007). *Night Haunts. A journey through London at night.* London: Verso.
Shaw, R. (2018). *The nocturnal city.* Abingdon: Routledge.
Sloane, M., Slater, D., & Entwistle, J. (2016). *Tackling social inequalities in public lighting.* Retrieved from: http://eprints.lse.ac.uk/66626/1/__lse.ac.uk_storage_LIBRARY_Secondary_libfile_shared_repository_Content_Sloane,%20M_Tackling%20Social%20inequalities_LSE-Tackling-Social-Inequalities-in-Public-Lighting-May-2016.pdf
Smith, A. (2016). *Events in the city: Using public spaces as event venues.* Abingdon: Routledge.
Stevens, A. (2016, June). How night mayors are boosting economies. *Cities Today.* Retrieved from https://cities-today.com/how-night-mayors-are-boosting-economies
Stevenson, D. (2018). Feminism and its places: Women, leisure and the night-time economy. In *The Palgrave handbook of feminism and sport, leisure and physical education* (pp. 557–569). London: Palgrave Macmillan.
Talbot, D. (2016). *Regulating the night: Race, culture and exclusion in the making of the night-time economy.* London: Routledge.
Thurnell-Read, T. (2012). What happens on tour: The premarital stag tour, homosocial bonding, and male friendship. *Men and Masculinities, 15*(3), 249–270.
Williams, R. (2008). Night spaces: Darkness, deterritorialization, and social control. *Space and Culture, 11*(4), 514–532.
Zukin, S. (1989). *Loft living: Culture and capital in urban change.* New Brunswick, NJ: Rutgers University Press.

http://orcid.org/0000-0002-9262-8983

Decoding middle-class protest against low-cost nocturnal tourism in Madrid

Begoña Aramayona and Rubén García-Sánchez

ABSTRACT

Nightlife scenes in central Madrid have been profoundly affected due to the expansion of tourist-oriented night-time leisure activities during the last three decades. This paper examines a range of tensions that have recently appeared in the La Latina neighbourhood due to the conflictual coexistence between the slow and locally-oriented everyday practices remaining in this territoir and the rapid colonisation of this central quarter of Madrid by neoliberal economies of the 'Tourist City'. Particularly, we focus on some long-term, middle-class residents who reproduce exclusionary narratives against the rapid expansion of low cost tourist-oriented nightlife, while advocating a civilised and distinctive tourism. We argue this may be seen as a renaissance of a sanitised 'middle-class culture' created by the fascist regime in the second half of the twentieth century. Recent middle-class' protests in Spain's largest cities hide a new struggle about 'who is legitimised' and 'who not' to reclaim the re-appropriation of the city centre.

Introduction

In May 2017, *The Sun*[1] newspaper published a piece describing the expulsion of a group of British tourists from a Ryanair flight to Alicante: 'WOMEN BEHAVING BADLY. "Vile" boozed-up Brit tourists hauled off Ryanair flight to Alicante by cops after knocking back vodka, swearing and acting like "creatures"' read the title. The online report was accompanied by a video in which five young women defiantly exited the plane, escorted by company employees, while passengers booed and celebrated their expulsion. Hours later, *The Sun* published a selfie of the women sunbathing once they reached their destination. 'Don't five girls have the right to have a good time on holiday?' asked one of them on her Twitter. The news did not take long to reach the Spanish press, which harshly criticised the antisocial behaviour of the tourists.

Neither the facts, nor the media narrative about them, are exclusive to the British press or Spanish beach destinations. The discomfort resulting from the increase in tourism in large Southern European cities has been widely publicised both in the press and via protests of various kinds (Colomb & Novy, 2017; Sequera & Nofre, 2018). In addition to the generation of precarious employment (Hernández, 2016; Santana, 2000) or the reduction of long-term housing arising from the emergence of tourist apartments (Arias & Quaglieri, 2016; Cócola, 2016), some residents lament the perceived impossibility of coexistence between residential and tourist leisure practices (Nofre et al., 2017a, 2017b). The development of 'drinkatainment' (Bell, 2007), resulting from the growth of the night-time economy (NTE) on a global scale, has triggered some of these protests, in which the visitors' 'uncivilised behaviour' contributes to the residents' discomfort.

Madrid's nightlife, the case study examined here, has also been profoundly affected by the expansion of tourist-oriented industries in the city over the last three decades. The inclusion of the central neighbourhood of La Latina in the nocturnal circuit has led to social tensions due to the conflicting coexistence of the slow and locally-oriented everyday practices preserved in this *territoir* and the rapid colonisation of this central quarter of Madrid by formal and informal neoliberal economies of the 'Tourist City' (Colomb & Novy, 2017). However, some of the arguments expressed by long-term residents against the increase in Madrid's nocturnal tourism seem to rely on class-based stereotypes that demonise the 'annoying' revellers and pose serious questions around 'who has the right' and 'who not', and 'who can legitimately' and 'who cannot' reclaim the city centre. In this sense, the question of 'whose emotions, sensations and experiences matter' (Hadfield, 2014, p. 7) or, as Sharon Zukin (2011) states, who exercises the 'moral property' of a place, seems important when considering who has the right (and who does not) to build the narrative of places that accumulate centuries of history. In sum, different levels of access and distribution of the 'politics of mobility' (Adey, 2006, Cresswell, 2010, Hall, 2005) in Southern European city centres are at the core of some of the recent and current struggles of the 'Tourist City' (Colomb and Novy, 2017).

This study analyses these tensions in light of the critical analysis of certain middle-class protests against the growth of low-cost nocturnal tourism in La Latina. Based on ethnographic fieldwork conducted between 2014 and 2016 – which included direct observation and 40 interviews with neighbours, revellers and business owners, as well as the analysis of social media information shared in different internet blogs created and administered by organised residents opposed to urban changes in the area (in total: 30 posts and associated comments; $N = 122$) – we aim to shed light on how the 'right to the city' (Lefrebve, 1968) might be appropriated and contested in diverse and sometimes conflicting ways by different inhabitants. In doing this, we engage with current debates around how Southern European middle-class culture is experiencing a profound economic, symbolic and cultural crisis in the aftermath of a deep recession and, secondly, how the social imaginaries of the night, leisure and tourism have shifted accordingly. In the first section, we analyse the social tensions arising from the increase in nightlife economies resulting from low-cost tourism in large Southern European cities. Second, we explore the changing nature of Madrid's nightlife scene from an historical perspective, and how a 'nyctophobic' culture (Edensor, 2013) or 'fear of the night' has been constructed alongside the city's political and economic changes since the second half of the twentieth century. In the third section, we specifically analyse how the narratives of some current long-term, middle-class residents against the expansion of low-cost tourist-oriented nightlife in La Latina produce and reproduce exclusionary narratives against class-based 'problematic bodies', while advocating for a civilised and distinctive tourism. We argue that such exclusionary narratives may be considered as a renaissance of the socially and politically sanitised 'middle-class culture' created by the fascist regime in the second half of the twentieth century, as well as the manifestation of the new shapes that the 'revanchist city' (MacLeod, 2002) acquires in the age of urban neoliberalism. In parallel, using the 'new mobilities paradigm' we aim to shed light on some of the conflicts underlying the current global geographies of tourism (Hall, 2005, Hannam et al., 2006, Sheller and Urry, 2006). Finally, we offer an alternative approach to the analysis of middle-class-led protests against the growth of contemporary low-cost tourism in Southern European urban areas, and propose that the 'right to the city' (Lefebvre, 1968, Harvey, 2008) may need to incorporate a postcolonial and feminist 'nocturnal citizenship' (Gwiazdzinski, 2014) to create fully inclusive nocturnal centres.

Nightlife and 'low-cost' tourism: tensions around the ludification of the 'tourist city' in Southern Europe

The impact of globalisation and migration on new types of night-time consumers (Hou, 2010; Pottie-Sherman and Hiebert, 2015) and the changing social representations of night and darkness (see Edensor, 2013) have been gaining attention in urban nightlife studies, although mainly focusing on an Anglo-Saxon context. The role of nightlife in revaluing and gentrifying the post-industrial city has also been extensively addressed (Hae, 2011, Nofre and Martín, 2009; Ocejo, 2014). Similarly, access to and inclusion in night-time centres (Eldridge and Roberts, 2008; Roberts and Eldridge, 2007) and the class-based, gender-based and race-based biases against less privileged potential night-time consumers in post-industrial cities have received critical attention (Eldridge, 2010; Thurnell-Read, 2013; Haydock, 2014). However, as we will argue throughout this paper, we have also witnessed a re-emergence of the 'moral panics' associated with 'alcohol culture' (Gee, 2014; Spracklen, 2014), as well as new forms of stigmatisation of 'problematic bodies' in public space (Eldridge, 2010; Thurnell-Read, 2013), associated with the intensification of internal and transnational low-cost nocturnal tourism in Southern European cities. At the same time, these biases are not new, and they respond to a long historical construction in countries with a past –and often present– colonial, patriarchal and elitist underlying culture.

The 'new mobility paradigm' plays a fundamental role in order to understand some of the deeper tensions that underlie the touristification of South European cities (Hall, 2005, Hannam et al., 2006, Sheller and Urry, 2006). Particularly concerning the night, new urban nocturnal enclaves and 'new mobile subjects' have appeared, such as a 'sanitised night' for Erasmus students (Malet, Nofre and Geraldes, 2016), new transnational nocturnal consumers (Malet, 2017; Nofre et al., 2017a) and nightlife venues for the European middle classes devalued by the crisis, all seeking to satisfy their leisure needs at affordable prices. As a result of the insertion of the global 'entertainment machine' (Clark, 2004; Lloyd and Clark, 2001) into Southern European urban centres, diverse 'constellations of mobilities' are now interacting in novel, sometimes conflicting, ways. Hence, as explained by Cresswell (2010), physical movements and their associated representations and mobile experiences are all part of a new struggle concerning the contemporary 'politics of mobility' on a global scale. Hence, in spite of Western narratives about 'mobility as progress' and the EU mantra of 'free movement', different mobilities are produced and productive of unequal social relations of power inside the European Union. Many 'party spaces' in a vast array of European cities are depicted and imagined as 'other places' where transgressive behaviour is permitted or even commodified by turning them into a brand for the visitor's consumption.[2] Examples of this complex relationship between commodification and transgression can be found in Thurnell-Read (2011), who discusses an interesting case in Poland, and Colomb (2012) who focuses on Berlin. However, due to the different impacts the recent crisis has had on the 'two-speed Europe', as well as its specific historical legacies, it seems that many Southern European cities are recently intensifying their role as 'ludic places' for the visitor's consumption, leading to diverse, contradictory and complex effects on the physical and socio-economical fabric of its cities.

In the Spanish case, some medium-sized cities such as Magaluf on the island of Majorca, but also big cities like Barcelona or Madrid, are beginning to experience an increase in visitors looking for a space to 'party' apparently without the rules and restrictions of the modern 'daily grindstone' of work and family life (Baptista, 2005). As authors such as Kelly, Hughes and Bellis (2014) have described in relation to Ibizan workers, the concept of 'work hard and party harder' has gained considerable currency. Both Spanish and international media have quickly echoed this new reality, with headlines describing the 'shamelessness' (Huffington Post, 2015) and 'ethyl debauchery' (El País, 2016) represented by the 'drunk tourism' phenomenon. The very content of protests by long-term residents in these places also incorporates the discomfort caused by the new low-cost 'drunk tourism', which they associate with the absence of 'civilised behaviour', an increase in crime, fighting and noise, as well as the serious difficulties of residential and new leisure activities coexisting (Nofre et al., 2017a, 2017b).

Nightlife in Madrid: new forms of commodification in the contemporary tourist city

There is a specific history about of the ways neoliberalisation happened in Madrid's nightlife. Madrid's nocturnal scene echoes the urban, residential, economic and cultural changes that have taken place in the city and its urban centre in recent decades. The long presence of a totalitarian fascist regime in Spain (1939–1975), led by Dictator *Generalisimo* Franco, left a National-Catholic morality ingrained in all spheres of public and private life – from respectable behaviour on the streets to sexuality in married life. The time after dark became the temporial and symbolic space of resistance to the imposition of a 'Francoist civility': late-evening cafés were the meeting place for clandestine political groups; cabarets, brothels and 'seedy' taverns, seeking to escape the harsh censorship of the regime, secretly appeared in Madrid's downtown. Paradoxically, although the complex political, military and ecclesiastical structure of the Spanish fascist regime allowed or even consumed in many of these spaces, a 'nyctophobic' narrative (Edensor, 2013) was constructed around the night and its 'Red' and 'ill-fated' sins, temptations and threats. To build 'a society of owners, not a society of proletarians' (Arrese, 1959), the new division between 'good and bad Spaniards' was carefully installed as the normative rationale in the neighbourhoods of Madrid. The 'good citizen' was exemplified by those who remained silent, worked hard and did not display 'rebellious behaviours'.

After Franco's death (1975), the emergence of neoliberal urban policies during the 1990s and 2000s by successive democratic governments and political and cultural inheritors of the fascist regime (López-Salinas, 2004; Maestre, 2013), was accompanied by a 'punitive turn' in relation to the night. In the 1980s, in a context of public disinvestment and urban deterioration of Madrid's urban core, a new ludic and political wave of counter-cultural scenes developed and supported by young people arrived and filled the urban space during both day and night. In contrast to previous decades, and although not exempt from police and extremist right-wing groups' harassments, a public and 'proud' narrative about marginal nightlife was deployed by these young groups. In central Madrid, the gay militant movement in the neighbourhood of Chueca, the underground, bohemian and punk music in Malasaña, and the squat, anarchist and left-oriented movements in Lavapiés coexisted with a strong presence of an older population. The narrative about 'degradation' was not long in coming. Increased by the emergence of heroin consumption and AIDS in the 1980s, the urban 'marginalization debacle' allowed the entry of the first rehabilitation policies, mostly financed with European funds during the 1990s. At the same time, local councillors like Angel Matanzo (popularly known as the 'Sheriff of Madrid') closed down dozens of premises, limited opening hours and organised raids that he himself directed in person against sex workers, drug users, street vendors and street musicians (the 'decadence of the heart of the city', in his words[3]). This 'social clean-up' set up the aesthetically 'hygienic' scenario needed for the promotion of Madrid as a 'global city' during the 2000s – a node for financial, business and service activities – and the intensification of the gentrification processes through public aid to private investors (García, 2014; Sequera, 2014; Sequera and Janoschka, 2015). In different stages, Madrid's diverse nocturnal counter-cultural scenes were replaced by other more 'sanitized' and 'daily-oriented' ones. For example, the 'dark rooms', saunas and other forms of furtive sexual practices of the stigmatised 'gay ghetto', were dispersed in favour of a concentration of more gleaming, liberal, middle-class 'gay friendly' cafes and restaurants; at the same time, the more feminine, trans and radical image of the gay movement during the 1980s was replaced by a virile and economically-successful representation of a 'gay in a suit' by media and event promoters during the mid 1990s and 2000s (René, 2016). Similarly, the dark and rebellious nightlife scenes in Malasaña were replaced by small art galleries, expensive vintage shops and 'veggie' restaurants for new upper-middle-class residents and consumers (Moreno-Ruiz, 2016).

Since 2000, the prohibition of a popular leisure practice in numerous cities, *Botellón*, added further pressure to the control of unregulated or non-commodified night-time practices. *Botellón*, common in Spanish cities and villages since the 1980s, is the informal and usually spontaneous gathering in small or large groups in public spaces –streets, squares, parks – accompanied by the consumption of alcoholic or non-alcoholic beverages. This enduring pastime encompasses an endless number of practices and generational profiles and it owes its existence to the opportunity to consume more economically compared with formal bars or venues and to the cultural idiosyncrasy of the Mediterranean context. In Madrid and many other Southern European areas, the combination of good Mediterranean weather and typically small-sized homes, has triggered the frequent use of public space throughout most of the year. However, since the 2000s, *Botellón* has been strongly stigmatised by the mass media, criminalised by policy and pathologised by non-critical academics as a representation of the 'excesses of youth'. In Madrid, the 2002 'Law on Drug Addiction and Other Addictive Disorders', popularly known as the '*Anti-Botellón Law*', officially regulated public alcohol consumption by prohibiting it in non-regulated establishments –only in bars, pubs or open-air terraces– and forbidding its sale in shops after 10pm, accompanied by increased policing and administrative sanctions for non-compliance. At the same time, since the 2000s, sensationalist reports of problems with 'dirt' and 'noise' have appeared in the media, as well as a great deal of scientific literature around 'violence' and 'public health' problems derived from *Botellón* (see Rodriguez-Martos, 2007; Basante, 2009), especially in relation to the 'vulnerability' of young people (see Calafat, 2007; Gómez-Fraguela et al., 2008). Following Spracklen (2014), we could say that this rhetoric about Spanish nightlife and *Botellón* exemplifies the paradox of neoliberal governments of the 21th century: the generation of policy-driven moral panics against 'alcohol culture' astonishingly coexists with an active impulse by the State, allied with alcohol production industries, to design marketable images around places in which alcohol plays a fundamental role; some even being transformed into a national brand for tourism promotion.[4] Hence, we see the active role of the Spanish neoliberal state in preparing the conditions for private accumulation processes (Brenner, Peck and Theodore, 2010) through encouraging alcohol consumption, while at the same time, paradoxically, criminalising it. However, the rationale of the 'civilised good citizen' (Haydock, 2014; Koning, Jaffe and Koster, 2015) that permeates these 'moral panics' is profoundly influenced by certain class-based, race-based and gender-based biases. Its construction responds to a long process of historical layering, which we analyse in the following section.

'They are beasts': genealogy of civilised 'middle-class culture' against 'low-cost' nocturnal tourism in La Latina's neighbourhood

The neighbourhood of La Latina, in the central district of Madrid, is popularly known as the most authentic, or *castizo*, neighbourhood in Madrid. A place of transition and folklore in Madrid since the seventeenth century, its current residents, mostly white middle-class since the 1990s,[5] coexist with the practices of alcohol-fuelled weekend crowds of international and Spanish tourists –the latter coming from the provinces near Madrid or from the city's working-class outskirts– that flock to the area every weekend, especially during the night. Due to its proximity to other tourist attractions in Madrid's city centre (such as *Palacio Real, Puerta del Sol* and *El Rastro*, the old flea market), La Latina's night-time economy boomed in 2008. The presence of bars and nightclubs soared on the streets of the Northern quarter (*Cava Alta* and *Cava Baja*), offering food and drinks at very low prices, until there were 450 such nightlife venues by 2016 (Madrid City Council Registry, 2016), most of them changing in short periods of time. The illicit licensing of nightlife activities, with a corruption network existing between business owners, local government officials, and municipal police, occurred alongside an increase in public terraces and police harassment of the already sanctionable '*Botellón*', which regularly took place on some of its streets. In 2014, some large clubs were already open on another major artery of the neighbourhood – the *Shoko* club on Toledo Street, for example– and, together with the promotion of the La Latina brand designed by event organisers and music festivals, consolidated the area's nocturnal economy. Notably, since the 1980s La Latina has been the central area of Madrid with the highest proportion of elderly residents (Madrid City Council Registry, 2017) whilst being simultaneously described by TripAdvisor (2016) as 'the best place to enjoy *cañas, tapas* and drinks in Madrid'.

The reaction by residents did not take long to surface. In 2006, a residents' association, mainly formed by the adult and elderly population, began to denounce the 'massification' of the area's NTE and the transformation of the place into an 'alcohol-fuelled neighbourhood', as well as the increased noise and dirt resulting from the opening of 'low-cost establishments', which only attracted 'low-cost people' (57-year old resident). Several websites were opened ('Residents against Shoko', 'Let's restore La Latina, Madrid of House of Austria') in which the loss of the neighbourhood's traditional identity as a result of overcrowding was also denounced. With no tangible changes, in 2016 residents went out on a Friday night 'for conviviality in La Latina'. Without intending to trivialise their discomfort, the headline of the banner they carried ('AREA CONTAMINATED ... by noise') conceals much of the complexity that we will analyse in the following paragraphs, relating to how some discourses of conviviality are permeated by colonial, patriarchal and classist traces that help to build an unequal scenario for the integral exercise of 'nocturnal citizenship' (Gwiazdzinski, 2014) by diverse, less privileged social actors.

The set of arguments against the 'overcrowding of low-cost people' and the threats of 'contamination' in this case are not accidental and transcend the noise issue. Multiple interconnected layers of identity based on gender, race, age and national culture were reclaimed when these organised neighbours rejected the massive influx of nocturnal tourists. Our focus here is the class-biased narratives underlying these residents' discourses, precisely because we consider it as one of the most inconspicuous elements behind this general 'revanchist' narrative, both inside the middle-class residents' discourse and in the general media narrative against the current working-class, 'uncivilised' and 'brutal' nocturnal tourists. Owen Jones (2011) has described how the discourse that 'demonises the working class', contemptuously referred to as 'chavs', and the 'war between the poor' constructed by the mass media and Thatcher's administration, has been aiding the implementation of the neoliberal project in England since the 1970s. In relation to nightlife, several authors have pointed out how 'binge drinking' culture in the British context has been actively linked with the 'excesses of the working class' (Hubbard and Colosi, 2015; Nayak, 2006) and reproduced class distinctions. As proposed by Hubbard and Colosi (2015), the construction of a middle-class historical subject that hegemonises 'civilised and respectable' subjectivity versus the 'gross and vulgar', also affects how the night and different types of leisure are perceived, promoted or hindered. In La Latina, the residents' rhetoric regarding their discomfort and the impossibility of coexistence also disguises this narrative. Wild and brutalised, turned into a homogenous mass, devoid of personality or diversity, the revellers become an unwanted object to be expelled from a physical space:

> Surreal news of the day: The crew of a Ryanair flight on the Liverpool-Alicante line expelled a group of British girls who would not stop drinking alcohol, shouting and behaving "like real animals", according to the English media. It would be great if we could do the same with the **savages** who come to the neighbourhood every weekend and behave like **animals**, wouldn't it? (Comment from a 57-year old resident on a website in relation to the news mentioned in the Introduction)

On the other hand, discourses of 'contamination' are common in colonial narratives based on the dichotomies of 'Us'/clean and 'the Others'/dirty (Douglas, 1966, Jodelet, 2007; Speltini and Passini, 2014), which we consider a fundamental component of the Franquoist legacy of the Spanish 'middle-class culture', has not yet been deeply explored. Purity is an innate symbolic feature of racial superiority in colonial discourses, and is often associated with the distinction of white and bourgeois people (Dussel, 2003; Mignolo, 2003). In that sense, the appeal to 'disgust' (Hubbard and Colosi, 2015; Lawler, 2005; Tyler, 2008) prompted by certain nocturnal practices seems to help set the scene for exercising a 'respectable' and 'decent' moral superiority over the excesses associated with the working classes. Certain types of establishments that seem to 'lower the tone' of the neighbourhood, such as *striptease bars* (Hubbard and Colosi, 2015) or *X Cinemas* and *Sex Shops* (Hubbard, Boydell, Crofts, Prior, & Searle 2013), but also certain uses of public space, such as hen parties (Eldridge and Roberts, 2009) or urinating in public (Eldridge, 2010), are described as 'amoral' and 'dirty', disregarded by the hegemony of the middle classes in the construction of night spaces. As Lawler argues, 'disgust is the manifestation of a bourgeois project to distinguish the middle class from its others' (2005, p.443). Under the scope of the 'new mobility paradigm', movements, representations and practices are entangled in complex and intimate ways to set diverse forms of 'constellations of mobilities' (Cresswell, 2010; Hannam et al., 2006; Sheller and Urry, 2006). Adding Mary Douglas' (1966) insights, we could argue that, based on the symbolic depiction of tourists, 'botellón' consumers and revellers as 'matter out of place', disgust might be the corporeal experience mobilised by these long-term residents when facing La Latina's urban change. The hierarchical order discursively installed by these middle-class residents might eventually result in the actual physical displacement of these annoying and 'dirty' subjects:

> As if we don't have enough people in the neighbourhood for San Isidro [annual traditional local celebration], there's been even more hype so that now it's completely packed … (…) What is supposed to be a neighbourhood celebration has turned into a **trashy macro-botellón party** … The further away they are from me, the better … **IT'S DISGUSTING.** (Comment by a 30-year old resident on a Facebook page)

As with the moral panics associated with *binge drinking* and the 'excesses of the working-class' (Nayak, 2006; Spracklen, 2014), moral panics around *Botellón* implied that their users –mostly, but not all, young working-class people– have been subjected to social, moral and policy-based surveillance. In the case of La Latina, the application of the 2002 *Anti-botellón* law against the uninhibited and low-cost practice of *Botellón* brought together organised residents and regulated nightlife businesses, both demanding more police patrols. However, this punitive or revanchist response against *Botellón* makes 'a brutalising demarcation of winners and losers' (MacLeod, 2002: 605) in which not only its consumers are affected. Around it there is also the informal or precarious employment of racialised migrants from the global South – mainly the Bangladeshi population lacking official paperwork and selling cans of beer for one euro, or the Chinese population which, bypassing the time restriction regulations, offers the purchase of alcoholic beverages – who find themselves in legal limbo. The unequal access and distribution of power in terms of a 'politics of mobility' (Adey, 2006, Cresswell, 2010, Hall, 2005) are again represented here. We see how the representation of these mobile migrants as 'swamping' and 'annoying' by Spanish local actors, residents and formal business owners who demand police action, impacts the practical and corporeal experience of these less powerful subjects. Paradoxically, these precarious workers are constantly forced to move (escaping from their country, escaping from Western police and deportation) while being economically stuck to place and having very restricted and surveilled movements.

In any case, the narrative of the genuine 'urban village' (Gans, 1962; Zukin, 2011) of La Latina idealised by these residents, compared with the low-cost night, hides a logic of exclusive *heritagisation* that, 'through its silences', (Hadfield, 2014: 4) builds a biased version of cultural identity. Despite the area's demographic variety and historical complexity,[6] the identity projected by this neighbourhood's Residents' Association over-represents its residential, white, heterosexual and middle-class family element: the epithet of Spanish civility. Following Cresswell (2010), the whole 'constellation of mobilities' reproduced by these middle-class residents seem to be sustained over the desire of preserving certain traditional values associated with 'parochialism' at a local scale (Tomaney, 2013) in contrast to a more 'global sense of place' (Massey, 1991, 2009) of a rapidly changing world. In contrast to the more mobile nomad subject embodied by the tourist, migrant or 'reveller', the biographical trajectories of these residents push them to be loyal to static and stable forms of conviviality represented by the rural or urban village. There are competing rights to the city at play in this case, not only between these middle-class residents and other social actors, but in the inner conceptions of the imagined neighbourhood by these 'annoyed' residents themselves. Indeed, as defended by Tomaney (2013) there is a powerful political potential embodied in these parochial demands of a less-mobile urban space: the importance of slow-built neighbourhood networks ('everyone knew and cared for each other') or the non-monetarized use of the public space ('my children could play freely in the street and we knew they would be fine'). However, at the same time, we argue here that the rationale of morality, respectability and disgust that underlies the discursive construction embodied by these middle-class residents represents a reproduction of the urban home-ownership class built during the second half of the twentieth century during Franco's fascist regime. As we argued before, the social meritocracy constructed during that period was based on the value of hard-work; it glorified *resting* –quiet and at home– and buried *laziness* –exhibited in the street–- as a right. The benefits of the 'hard hand' and an image of 'peace, order and cleanliness', with strong allusions to the splendour of the colonial past, were partially installed as a general social rationale. These 'reactionary politics' (Massey 1991, 2009) inherited as a Francoist legacy, might be also at the heart of the anti-mobility neighbour's demands, and appears consistently in middle-class residents' arguments against the 'low-cost night': disdain for all that smells of the popular or subaltern, the search for symbolic centrality – the final result of the rise in social status exemplified by the middle classes – and the obsession with planning and policing as a way to avoid the overspill of any peripheral movement. In this case, the rejection of *Botellón,* as a singular and uninhibited use of public space and the 'noisy' and 'uncontrolled' use of La Latina's 'overcrowded' night, is symptomatic of this rebirth of a sanitised heritage that is articulated in many of today's Spanish urban middle-class owners.

Final remarks

In the case of La Latina, protests by long-term residents against the growth of low-cost nightlife hide elitist and segregating narratives in which the perceived need to defend genuine 'traditional residents' is used to generate a coherent internal identity against working-class nightlife and its users. The use of the argument for a genuine 'urban village' contains underlying potential against the commodification and hyper-regulationist practices of the urban space, but it also hides a worrying logic of exclusive heritagisation of the place that renders invisible any alternative and more radical narratives of contemporary urban change. Thus, a critical analysis of the rejection of low-cost culture by these Spanish middle-class residents, represented in this case through the rejection of the *Botellón* and the 'low-cost night', can be useful for understanding some of the recent debates surrounding the growth of urban tourism, different typologies of protest, and non-neoliberal alternatives of urban change in South European large cities. Although we base our understanding of these concealed tensions on a class-based approach, future research might address the diverse and complex intersections between different forms of identity such as class, gender, race, age and national culture.

The Lefebvrian 'Right to the City' (1968) or the 'Right to Place' is becoming a motive for local urban opposition to the intensification of neoliberal policies in Southern Europe since 2008. The emergence of mass, low cost tourism poses new challenges for the analysis and interventions of urban centres experiencing a growing process of 'ludification' (Baptista, 2005). Although the purpose of this study was not to address the effects and impact of this new wave of resident's contestation, we hypothesise that there is high political potential for urban mobilisation that may affect other areas experiencing urban conflict. Perhaps what the social context of La Latina demonstrates is the complexity of the ways in which the 'right to stay put' can be used by diverse actors, as well as the need to revisit some of the arguments put forward that seem to respond to certain problematic colonial and patriarchal cultural legacies, as well as the 'monopoly of true humanity' (Lawler, 2005) that embodies some middle-class protests in post-industrial cities. The enduring practice of *Botellón* in large Spanish cities is an example of how much another approach to the use of night spaces is needed. At the same time, resistance to the 'nyctophobia' (Edensor, 2013) constructed for decades in Western cultures and the uninhibited enjoyment of the night may help generate a new nocturnal citizenship (Gwiazdzinski, 2014) that affords the night, together with the 'disrespectful', 'gross' and 'distasteful', the legitimacy they deserve. In this sense, the use of feminist and postcolonial perspectives would be helpful when critically examining the analysis of resident protests against the recent growth of tourism-oriented industries in Southern Europe.

Notes

1. https://www.thesun.co.uk/news/3622179/ryanair-hen-party-alicante-flight/.
2. We appreciate this comment made by one of the reviewers of a previous draft of this work.
3. https://elpais.com/ccaa/2017/03/06/madrid/1488834215_666182.html.
4. In 2012, the design of the 'Spain brand' emphasised the importance of enhancing the nightlife economy as a distinctive element of Spanish tourism marketing. Others, such as the image of the popular 'Osborne' cognac and its famous 'bull', have been appropriated as patriotic symbols since the 50s and 60s until today.
5. Several Plans for urban renovation, particularly of public spaces and residential housing, were developed in this area during 1994–2003 and 2008–2011, cementing the neighbourhood as a middle-class, marginal-free area (Aramayona, García-Sánchez, Martín, Martínez, & Corraliza, 2019)
6. The constant allusion to the exclusively 'historical' character of the neighbourhood, with special emphasis on its monarchical, noble and imperial past ('Madrid, the village of the House of Austria', 'the ancient and noble area of Madrid') renders other residential

population profiles invisible and fails to recognise the variety of people who, *de facto*, inhabit the area although without being residents or home-owners: although marginal, low-income people still live there (with differences in household income of up to 25,000 euros per year). The 'noble district' narrative also renders invisible other events that could form part of the historical narrative of the place, such as the anti-fascist political struggles during the Spanish Civil War (1936–1939) or the clandestine left-wing struggles during Franco's regime, especially strong during the Autarchy period (1939–1959), as well as other marginal practices such as the high presence of elderly women working as prostitutes on its streets during the fifties and sixties or the emergence of gay venues popularly known for allowing practices such as urinating or nudity until the nineties.

Disclosure statement

No potential conflict of interest was reported by the author.

ORCID

Begoña Aramayona http://orcid.org/0000-0002-4687-8500
Rubén García-Sánchez http://orcid.org/0000-0002-0472-5717

References

2002/5 'Law on Drug Addiction and Other Addictive Dissorders'. BOCM.

Adey, P. (2006). If mobility is everything then it is nothing: Towards a relational politics of (im)mobilities. *Mobilities, 1*(1), 75–94.

Aramayona, B., García-Sánchez, R., Martín, M. J., Martínez, J. M., & Corraliza, J. A. (2019). ¿Vecinos de toda la vida? Nimby, ocio nocturno y desapropiación en centros urbanos. *La Latina, Madrid. Athenea Digital, 19*(1), 1–27.

Arias, A., & Quaglieri, A. (2016). Unravelling Aribnb: Urban perspectives from Barcelona. In A. P. Russo and G. Richards (Eds.), *Reinventing the local in tourism: Producing, consuming and negotiating place* (pp. 209–228). Bristol: Channel View Publications.

Arrese, J. L. (1959). *No queremos un país de propietarios sino de proletarios.* ABC Press.

Baptista, L. V. (2005). Territórios lúdicos (e o que torna lúdico um território): Ensaiando um ponto de partida. Actas dos ateliers do V° Congresso Português de Sociologia. Sociedades Contemporâneas: Reflexividade e Acção Atelier: Cidades, Campos e Territórios.

Basante, A. R. (2009). Spain. In P. Hadfield (Ed.), *Nightlife and crime: Social order and governance in international perspective* (pp. 153–165). Oxford: Oxford University Press.

Bell, D. (2007). The hospitable city: Social relations in commercial spaces. *Progress in Human Geography, 31*(1), 7–22.

Brenner, N., Peck, J., & Theodore, N. (2010). Variegated neoliberalization: Geographies, modalities, pathways. *Global Networks, 10*(2), 182–222.

Calafat, A. (2007). El abuso de alcohol de los jóvenes en España. *Adicciones, 19*(3), 217–224.

Clark, T. N. (2004). *The city as entertainment machine.* New York: Roman and Littlefield Publishers.

Cócola, A. (2016). Apartamentos turísticos, hoteles y desplazamiento de población. Informe para el debate sobre el nuevo Plan Especial Urbanístico de Regulación de los Alojamientos Turísticos.

Colomb, C. (2012). Pushing the Urban frontier: Temporary uses of space, city marketing, and the creative city discourse in 2000S Berlin. *Journal of Urban Affairs, 34*(2), 131–152.

Colomb, C., & Novy, J. (2017). *Protest and resistance in the tourist city.* New York: Routledge.

Cresswell, T. (2010). Towards a politics of mobility. *Environment and Planning D: Society and Space, 28*, 17–31.

Douglas, M. (1966). *Purity and danger: An analysis of concepts of pollution and taboo*. New York: Praeger.
Dussel, E. (2003). Europa, modernidad y eurocentrismo. In E. Lander (Ed.), *La colonialidad del saber: eurocentrismo y ciencias sociales* (pp. 41–53). Buenos Aires: CLACSO-UNESCO.
Edensor, T. (2013). The gloomy city. Rethinking the relationship between light and dark. *Urban Studies*, 1–17. doi:10.1177/0042098013504009.
Eldridge, A. (2010). Public panics: Problematic bodies in social space. *Emotion, Space and Society*, *3*, 40–44.
Eldridge, A., & Roberts, M. (2008). A comfortable night out? Alcohol, drunkenness and inclusive town centres. *Area*, *40*(3), 365–374.
Eldridge, A., & Roberts, M. (2009). Hen parties: Bonding or brawling? *Drugs: Education, Prevention and Policy*, *15*(3), 323–328.
Gans, H. J. (1962). *The Urban villagers: Group and class in the life of Italian-Americans*. Nueva York: The Free Press.
García, E. (2014). Gentrificación en Madrid: de la burbuja a la crisis. *Revista de Geografía Norte Grande*, *58*, 71–91.
Gee, S. (2014). Sport and alcohol consumption as a neoteric moral panic in New Zealand: Context, voices and control. *Journal of Policy Research in Tourism, Leisure and Events*, *6*(2), 153–171.
Gómez-Fraguela, J. A., Fernández, N., Romero, E., & Luengo, A. (2008). El botellón y el consumo de alcohol y otras drogas en la juventud. *Psicothema*, *20*(2), 211–217.
Gwiazdzinski, L. (2014). Pleading for the right to the city's night. Night manifesto. Seeking citizenship 24h. *Invisiveis Produçoes*, 204–219.
Hadfield, P. (2014). The night-time city. Four modes of exclusion: Reflections on the Urban studies special collection. *Urban Studies*, *52*(3), 606–616.
Hae, L. (2011). Dilemmas of the nightlife Fix: Postindustrialisation and the gentrification of nightlife in New York city. *Urban Studies*, *48*(16), 3449–3465.
Hall, C. (2005). Reconsidering the geography of tourism and contemporary mobility. *Geographical Research*, *43*(2), 125–139.
Hannam, K., Sheller, M., & Urry, J. (2006). Mobilities, immobilities and moorings. *Mobilities*, *1*(1), 1–22.
Harvey, D. (2008). The right to the city. *New Left Review*, *53*, 23–40.
Haydock, W. (2014). The 'civilising' effect of a 'balanced' night-time economy for 'better people': Class and the cosmopolitan limit in the consumption and regulation of alcohol in Bournemouth. *Journal of Policy Research in Tourism, Leisure and Events*, *6*(2), 172–185.
Hernández, V. C. (2016). Prefordismo, fordismo y postfordismo en el turismo: Modelos para el análisis del turismo. *Revista Turydes: Turismo y Desarrollo*, *21*, 1–20.
Hou, J. (2010). 'Night market' in Seattle: Community eventscape and the reconstruction of public space. In J. Hou (Ed.), *Insurgent public space: Guerrilla Urbanism and the remaking of contemporary cities* (pp. 111–122). New York: Routledge.
Hubbard, P., Boydell, S., Crofts, P., Prior, J., & Searle, G. (2013). Noxious neighbors? Interrogating the impacts of sex premises in residential areas. *Environmental and Planning A*, *45*(1), 126–141.
Hubbard, P., & Colosi, R. (2015). Respectability, morality and disgust in the night-time economy: Exploring reactions to 'lap dance' clubs in England and Wales. *The Sociological Review*, *63*, 782–800.
Jodelet, D. (2007). Imaginaires erotiques de l'hygiène femenine intime. Approche anthropologique. *Connexions*, *87*, 105–127.
Jones, O. (2011). *Chavs. La demonización de la clase obrera*. Madrid: Capitan Swing.
Kelly, D., Hughes, K., & Bellis, M. (2014). Work hard, party harder: Drug use and sexual behaviour in young British casual workers in Ibiza, Spain. *International Journal of Environmental Research and Public Health*, *11*, 10051–10061.
Koning, A., Jaffe, R., & Koster, M. (2015). Citizenship agendas in and beyond the nation-state: (en)countering framings of the good citizen. *Citizenship Studies*, *19*(2), 121–127.
Lawler, S. (2005). 'Disgusted subjects': The making of middle-class identities. *The Sociological Review*, *53*(3), 429–446.

Lefrebve, H. (1968). *Le Droit à la Ville* [The Right to the City]. (2nd ed.). Paris: Anthropos.
Lloyd, R., & Clark, T. (2001). The city as an entertainment machine. In *Critical Perspectives on Urban Redevelopment. Published* (p. 357–378).
López-Salinas, A. (2004). *Notas de un testigo: la transición española inacabada*. Revista Unidad Cívica por la República. Retrieved from http://www.unidadcivicaporlarepublica.es/opinion2/transicion%20armando.htm
MacLeod, G. (2002). From urban entrepreneurialism to a 'revanchist city'? On the spatial injustices of Glasgow's renaissance. *Antipode*, *34*(3), 602–624.
Maestre, A. (2013, October 06). La mayoría silenciosa, el PP y el franquismo sociológico. La Marea. Retrieved from http://www.lamarea.com/2013/10/06/la-mayoria-silenciosa-el-pp-y-el-franquismo-sociologico/
Malet, D. (2017). Understanding international students beyond studentification: A new class of transnational urban consumers. The example of Erasmus students in Lisbon (Portugal). *Urban Studies*, *55*(10), 2142–2158.
Malet, D., Nofre, J., & Geraldes, M. (2016). The Erasmus Corner: Place-making of a sanitised nightlife spot in the Bairro Alto (Lisbon, Portugal). *Leisure Studies*, *36*(6), 778–792.
Massey, D. (1991, June 24–28). A global sense of place. *Marxism Today*.
Massey, D. (2009). The possibilities of a politics of place beyond place? *Scottish Geographical Journal*, *11*(1), 33–48.
Mignolo, W. (2003). *Historias locales/diseños globales. Colonialidad, conocimientos subalternos y pensamiento fronterizo*. Madrid: Akal.
Moreno-Ruiz, J. L. (2016). *La movida modernosa. Crónica de una imbecilidad política*. Madrid: La Felguera.
Nayak, A. (2006). Displaced masculinities: Chavs, youth and class in the post-industrial city. *Sociology*, *40*(5), 813–831.
Nofre, J., & Martín, J. (2009). Ocio nocturno, gentrificación y distinción social en el centro histórico de Sarajevo. *Anales de Geografía*, *29*(1), 91–110.
Nofre, J., Giordano, E., Eldridge, A., Martins, J., & Sequera, J. (2017b). Tourism, nightlife and planning: Challenges and opportunities for community liveability in La Barceloneta. *Tourism Geographies*, *20*(3), 377–396.
Nofre, J., Malet-Calvo, D., Geraldes, M., Sánchez-Fuarros, J., Martins, C., Pereira, P., … López, A. (2017a). Exploring nightlife and Urban change in Bairro Alto, Lisbon. *City & Community*, *16*(3), 330–344.
Ocejo, R. (2014). *Upscaling downtown: From Bowery Saloons to Cocktail Bars in New York city*. New York: Princetone University Press.
Pottie-Sherman, Y., & Hiebert, D. (2015). Authenticity with a bang: Exploring suburban culture and migration through the new phenomenon of the Richmond night market. *Urban Studies*, *52*(3), 538–554.
René, R. (2016). From guetto to gay village. Chueca in the media (1960-2000). *Espacialidades*, *6*(1), 105–140.
Roberts, M., & Eldridge, A. (2007). Quieter, safer, cheaper: Planning for a more inclusive evening and night-time economy. *Planning, Practice & Research*, *22*(2), 253–266.
Rodriguez-Martos, A. (2007). ¿Por qué es tan difícil legislar sobre alcohol en España? *Adicciones*, *19* (4), 325–332.
Santana, M. A. (2000). Turismo fordista y post-fordista: un aproximación a la operativización de conceptos. In M. Latiesa, & A. Álvarez (Eds.), *El turismo en la sociedad contemporánea: diversificación, competitividad y desarrollo* (pp. 72–95). Proyecto Sur: Granada.
Sequera, J. (2014). City, public space and neoliberal gobernmentality. *Urban*, *7*, 69–82.
Sequera, J., & Janoschka, M. (2015). Gentrification dispositifs in the historic centre of Madrid: A reconsideration of urban governmentality and state-led urban regeneration. In L. Lees, H. Bang Shin, & E. López-Morales (Eds.), *Global gentrifications: Uneven development and displacement* (pp. 375–394). Bristol: Routledge.

Sequera, J., & Nofre, J. (2018). Urban activism and touristification in Southern Europe: Barcelona, Madrid and Lisbon. In J. Obrahim, & J. Roberts (Eds.), *Contemporary left wing activism: Democracy, participation and dissent in a global context* (pp. 88–105). New York: Routledge.

Sheller, M., & Urry, J. (2006). The new mobilities paradigm. *Environment and Planning A*, *38*, 207–226.

Speltini, G., & Passini, S. (2014). Cleanliness/dirtiness, purity/impurity as social and psychological issues. *Culture & Psychology*, *20*(2), 203–219.

Spracklen, K. (2014). Moral panics or the politics of pleasure? Alcohol and policy directions in tourism, leisure and events. *Journal of Policy Research in Tourism, Leisure and Events*, *6*(2), 93–94.

Thurnell-Read, T. (2011). Off the leash and out of control: Masculinities and embodiment in Eastern European stag tourism. *Sociology*, *45*(6), 977–991.

Thurnell-Read, T. (2013). 'Yobs' and 'snobs': Enbodying drink and the problematic male drinking body. *Sociological Research Online*, *18*(2), 1–10.

Tomaney, J. (2013). Paroquialism: A defense. *Progress in Human Geography*, *37*(5), 658–672.

Tyler, I. (2008). Chav mum chav scum. *Class Disgust in Contemporary Britain. Feminist Media Studies*, *8*(1), 7–34.

Zukin, S. (2011). *Naked city*. Nueva York: Oxford University Press.

Gentrification, tourism and the night-time economy in Budapest's district VII – the role of regulation in a post-socialist context

Gergely Olt, Melanie Kay Smith, Adrienne Csizmady and Ivett Sziva

ABSTRACT

The processes of gentrification and tourism are often inextricably linked, but their relationship is not unequivocal: tourists can be explorers of stigmatised areas before gentrification or touristification can expel middle-class gentrifiers. In the post-socialist context, tourism and foreign consumers play an extremely important role in the gentrification process. This is especially true in our case study area the 'party quarter' in District VII of Budapest. The authors explore the inter-relationship between gentrification, tourism and the night-time economy in this area focusing on the effects of regulations and political struggles under post-socialist neo-patrimonial governance.

Introduction

In many cities around the world the growth of the night time economy (NTE) and tourism (which is often related to increased NTE consumption) are driving (re)investment, property price increases and displacement in particular neighbourhoods (Cócola Gant, 2016; Gotham, 2005; Gravari-Barbas & Guinand, 2017). In certain cases, reinvestment in the leisure economy leads to residential gentrification (Hae, 2011). In other cases however, the residential function of the neighbourhood is becoming less important (Islam, 2005; Pinkster & Boterman, 2017). Entrepreneurial and neoliberal city governance are widely used frameworks used to explain urban changes (Hubbard, 2017; Nofre, Giordano, Eldridge, Martins, & Sequera, 2017; Novy, 2017), however differences in the outcomes of urban change indicate the need to look beyond universal political-economic mechanisms (Barnett, 2010, p. 269). Urban policy and gentrification are not simply driven by entrepreneurial interests. Other, non-economic motives can be found behind them (Bernt, 2012, p. 3046).

Post-socialist cities are good examples of why 'gentrification … is in a way similar … but also diverges' from (western) cities where the phenomenon was first described (Bernt, 2016a, p. 641). There is a lively discourse today about the difference between post-socialist and other cities of the world (Tuvikene, 2016). Instead of looking at post-socialist cities as entirely different ('as a totality' (ibid. p. 141)) we can understand them in a de-territorialised way, by focusing on particular phenomena and processes. As a result of the post-socialist social and economic transformation we can observe a set of 'continuities' and 'anti-continuities' (ibid.) of socialism in certain institutions. The built form of a city or the authoritarian tendencies of public institutions are examples of the former while the imperative in governance to attract private investment and EU funds above all and 'state-phobia' (ibid.) are examples of the latter.

In the case of gentrification, because of the absence of solvent demand for high status inner city housing (Smith, 1996, p. 175), tourism and temporary foreign residents played a leading role in the reinvestment processes of many post-socialist cities (Sykora, 1993, 2005, pp. 93–94). The special importance of the NTE and tourism in relation to gentrification can be explained by the social and political contexts of post-socialist cities. However, as we will show, the local regulations largely determined how exactly the transformation played out.

After the end of socialism in 1989 'neo-patrimonial' power and property relations (Szelényi & Csillag, 2015, p. 26) influenced the mechanisms of urban change in Budapest. In neo-patrimonial contexts, besides the normal workings of market capitalism political power is to a large extent in a 'lord-vassal' relationship with economic players where traditional authority co-exists with the liberal (legal rational) authority. The result is policies with partisan or ideological motives and documented corruption in the allocation of public spending (Jávor & Jancsics, 2016). This has been especially true in Hungary since the 'illiberal turn' (Szelényi & Csillag, 2015) in 2010, after the 2/3 super majority electoral win of the right-wing party Fidesz, a result repeated in 2014 and 2018. Viktor Orbán, the current Hungarian Prime Minister and leader of the Fidesz defined his regime as 'illiberal'. In illiberalism the separation of powers, the defining feature of liberal democracy, is eroded and the executive branch dominates the judiciary and the legislature (ibid. p. 24–25).

Our aim is to show that institutional peculiarities influence how the NTE and tourism affect gentrification. Under legal rational (liberal) regimes local governments can be pressured to restrict tourism (e.g. anti-STR[1] measures in Berlin)[2] or left wing movements can win elections by appealing to anti-tourism concerns (e.g. in Barcelona[3]). The actual effectiveness of these policies and the realisation of campaign slogans can of course be questioned[4], but the issue certainly became part of the political discourse in these examples. However, in the absence of politically and legally accountable urban planning and policy for the NTE and tourism, it is difficult to enlist support from politicians. Even promises of politicians representing anti-tourism movements can be missing, as we demonstrate in our case study. Therefore, in this discussion we will focus on the regulatory context and the roles of particular interest groups. Before that, we will provide a brief literature review and outline the context of Budapest. Our conclusions summarise why the distinction between different political and regulatory systems can be important for the growth and management of both the NTE and tourism.

Leisure economy, gentrification and neoliberalism

Tourism creates potentially higher rents in urban areas (Gotham, 2005, p. 1103). The rent gap expands not simply because of disinvestment (Smith, 1979) but because of the functional change of the built environment (Wachsmuth & Weisler, 2017, p. 7). Beyond automatisms of capitalist disinvestment-reinvestment cycles (Smith, 1982, p. 151) enforced by neoliberal governance (Slater, 2017, pp. 89–90), demand side factors of post-industrialisation (Smet, 2016, p. 502) also play an important role in gentrification (see also Hamnett, 2003).

Looking beyond the automatism of the rent gap theory can explain why disinvestment was not followed by reinvestment for decades in the institutional context of Budapest (Czirfusz, Horváth, Jelinek, Pósfai, & Szabó, 2015, p. 58), and why it was necessary to apply conscious policies of promoting post-industrial functional change – for example tourism and the NTE – in many disinvested urban contexts (Roberts, 2009; Novy, 2017, p. 57). In property markets like New York, these policies led to residential gentrification and the pioneering leisure economy was displaced (Hae, 2011, p. 3460; see also Evans, 2017 for the case of London). In other cases, however, either the leisure economy remains determinant (Islam, 2005, p. 129) and can even displace high status owner occupiers (Pinkster & Boterman, 2017, p. 468) or tourism does not lead to instant residential gentrification (Smith & Pappalepore, 2015, p. 111). In certain contexts, the promotion of tourism and the NTE is seen as the reason for urban social tensions in housing as well (Cócola Gant, 2016; Novy, 2017), while in others, it is precisely the restriction of the NTE and tourism that leads to gentrification (Roberts, Turner, Greenfield, & Osborn, 2006, p. 1116; see also Roberts, 2009, p. 499). In short, the effects of functional changes on the housing market are dependent on the social and regulatory context.

The neoliberal framework is not necessarily sufficient to describe the logic behind urban policies in general, particularly in our case. For example, it can be argued that urban policies can also be motivated by 'recreat[ing] social order' (Bernt, 2012, p. 3046) beyond real estate market interests. The 'class project' (Harvey, 2005 cited by Barnett, 2010, p. 270) of neoliberalism restricts political power as a servant of capitalist interests. However, neoliberalism is only one governmentality among many (Ong, 2007), and in certain cases politics 'is still in command' (Szelényi, 2015, p. 48). Many use neoliberalism as the cause of everything else (Ferguson, 2010, p. 171) and take contextual differences as 'varieties of a single genus' (Barnett, 2005, p. 8). Others understand them as analytically different factors (Collier, 2012, p. 189). Either way, according to our research experience, the local institutional context needs closer examination to understand the actual mechanisms at play.

In certain processes and mechanisms (but not 'as a totality': see Tuvikene, 2016 p. 141) Budapest shows marked differences compared to cities with a more established neoliberal NTE-tourism agenda (such as Barcelona, Lisbon and Berlin – for example see Nofre, Giordano, et al., 2017; Nofre, Sanchez-Fuarros, et al., 2017; Novy, 2017). On the other hand, legal and political contestation can be more apparent in cities where the social context stands closer to liberal democracy.

Contextual differences and the role of the political context and regulation

In the post-socialist context, political power often functions in a neo-patrimonial manner (Szelényi & Csillag, 2015). Neo-patrimonialism (or 'feudalism' see Szelényi, 2016, pp. 14–15) means here that, during the political transformation and privatisation after the end of socialism in 1989, political leaders granted 'fiefs' for their 'vassals' (Szelényi & Csillag,

2015, p. 26; see also Stark & Vedres, 2012) to eventually strengthen their own power instead of enforcing market competition. In neo-patrimonial governance, traditional authority co-exists with legal rational authority (or liberalism) (Szelényi, 2016, pp. 7–10) and in illiberal regimes the executive branch of power dominates the legislative and judicial systems without the working of 'checks and balances'. This is especially true with a one party supermajority in parliament, hence the term 'illiberal democracy' (Szelényi & Csillag, 2015, p. 24). Traditional authority means unpredictably changing regulations often serving the temporary interests of certain political and economic groups, and less of a consistent class project of neoliberalism and market capitalism. The enforcement of existing rules is also selective.

The transformation of property rights, regulation of the NTE and urban conflicts are largely determined by the post-socialist political context. The differences between the ways of privatising housing in post-socialist cities (Bernt, Gentile, & Marcinczak, 2015; Sykora, 2005) resulted in markedly different transformation processes. For example, the pace of urban transformation was quicker in Berlin than in St. Petersburg (Bernt, 2016b, p. 577, 576) and Budapest (Kovács, Wiessner, & Zischner, 2013) thanks to the different institutional frameworks. In Budapest the neo-patrimonial way of housing privatisation and the fragmented ownership structure was an institutional setting that limited gentrification (Czirfusz et al., 2015, p. 58; Kovács et al., 2013, p. 5, 11) (see the details below). Meanwhile in Prague, for example, the restitution of whole buildings to former owners led to quick functional changes in the inner city (Sykora, 1993).

We can also observe the importance of the institutional setting in regards to the development of the NTE and tourism. Instead of the planned neoliberal boosterism witnessed in Berlin, (Novy, 2017), in Bucharest (Bürkner and Totelecan, 2018) and Budapest (Smith et al., 2018) a more unplanned commercialisation of certain inner city neighbourhoods could be observed. There are also contextual features in the handling of conflicts caused by the transformation. Cooperation and self-regulation of entrepreneurs (see for example Roberts et al., 2006, p. 1120) and their cooperation with the authorities (Hadfield, Lister, & Traynor, 2009, p. 474) is less likely in neo-patrimonial post-socialist states. On the other hand, residential struggles are also less effective in the post-socialist context (Kębłowski & Van Criekingen, 2014). In our case study we explain this in reference to the illiberal and neo-patrimonial governance. Instead of democratic debates over neoliberal planning we see political discourses against civil activism, creating rules according to the interests of 'vassals', and anti-entrepreneurial measures to appeal to the conservative 'law and order' voters in Budapest.

Methodology

This paper is based on ethnographic field research which started in 2006. Non-participant observation was carried out to follow the visible changes in the housing stock, retail offer, the use of public spaces and the prevalence of street crime, such as drug dealing and prostitution. Field observations concerning the physical changes of the neighbourhood, as well as the use of the streets and bars were noted and discussed with colleagues involved in the research. Over 50 interviews with residents and other stakeholders were recorded and 23 interviews most related to the topic of this paper are used here. Interviewees were selected after field observations (e.g. which buildings were the most affected by noise or by municipal interventions), by snowball sampling and from residents' forums. Several residents' meetings were attended and demonstrations of residents were observed. Four additional interviews were undertaken with members of a new residential movement which started in 2017. Between 2010 and 2015, eight local pub owners were also interviewed, some of them several times. They were selected according to their engagement in political struggles and their availability. Interviews with 3 real estate market experts working in the neighbourhood in 2011 and 2016 were also undertaken. One of the researchers also participated in regular meetings of the group of pub owners (called 'Azért') between 2010 and 2014. The interviews were recorded and transcribed.

The original research question of the ethnography back in 2006 was to see how planned urban rehabilitation affects the life of local residents. However, the transformation took an unexpected turn and we identified and followed the emerging conflict between a NTE-tourism cluster and residents. Examining our research results and the narratives of the stakeholders mentioned above from this perspective showed tourism development to be unplanned and even unintended in our case compared to the conscious touristification of cities such as Barcelona, Lisbon or Berlin (Nofre, Giordano, et al., 2017; Nofre, Sanchez-Fuarros, et al., 2017; Novy, 2017). Meanwhile, we found there was also less opportunity for residents to contest these processes.

The transformation of post-socialist Budapest and the case study area

Unlike in other post-socialist cities (Sykora, 2005), in Budapest the privatisation of state-owned housing after the 1989 regime change was carried out exclusively in a neo-patrimonial right-to-buy scheme for the sitting tenants who paid less than 10% of the actual market price for their apartments. By the late 1990s, 80% of the inner city housing was owner-occupied (Hegedűs & Tosics, 1998). The refurbishment costs of nineteenth century dilapidated buildings were delegated to the new owners who often lacked the financial means necessary to repair them (Kovács et al., 2013, p. 5), and the fragmented ownership structure of the buildings made investments difficult (Smith, 1996, p. 176). In 2010, Neil Smith expressed his surprise at the slow pace of gentrification in Budapest caused by 'institutional issues' (Czirfusz et al., 2015, p. 58). Functional gaps (Sykora, 1993) were also closed later than in Prague where many tenement houses lost their housing function and tourism played a major role in urban change (ibid. p. 288, 290). This was less typical in Budapest in the 1990s and 2000s. Gentrification accelerated substantially only after 2014 because of touristification and commodification of housing. This was fuelled by the macroeconomic trend of constant decrease of interest rates by the National Bank of Hungary (Csizmady, Hegedűs, & Nagy, 2017). These factors made real estate an attractive investment option and real estate prices increased by 40% in the inner city by 2016 (FHB, 2016).

Designated urban rehabilitation areas, where local authorities planned interventions in the urban fabric, were exceptions to the privatisation process. However, instead of a city wide programme, sovereign districts implemented different types of urban rehabilitation (Kovacs, Wiessner, & Zischner, 2015, pp. 258–259).[5] Our case study area in District VII is the so-called Inner-Elizabeth Town (Erzsébetváros), which is the most central part of the district, and the area of the WWII ghetto of Budapest. Because of the urban rehabilitation plan called Madách-promenade (Román, 1998), many buildings remained in the ownership of the local authority. From the early 2000s, these dwellings were vacated by the local authority and whole buildings were privatised, which created a very different background for urban change than flat-by-flat privatisation (see above). However, this process was carried out in a way which was reported by the Hungarian judicial system to be fraudulent[6] (see Sipos & Zolnay, 2009). This example of neo-patrimonial governance slowed down the transformation process (Csanádi, Csizmady, & Olt, 2012).

The emergence of ruin bars and tourism

The corrupt privatisation and the demolition of many heritage buildings in spite of the rules (see Sipos & Zolnay, 2009) were also typical neo-patrimonial elements of urban change in Budapest. During the slow privatisation process, the local authority gave permits to open summer terraces in the inner courtyards of the now vacated and dilapidated buildings, and the press named them 'ruin bars' (Földes, 2003). The bars were a great success, but before the 2006 local elections the local authority restricted licensing. The next boom of ruin bars started after the 2008 crisis when empty buildings, now privatised and owned by real estate speculators, had to produce some kind of income, after property values dropped by 20–25%. This time, rent contracts were much longer term, contradicting the prediction of temporary functional change (Lugosi, Bell, & Lugosi, 2010, p. 3) and national level licensing regulation became much looser in 2009 (see the details below).

Today the area is commonly referred to as the 'party quarter' although in 2011 there were still more than twenty thousand people living there. Hundreds of smaller and many larger venues operate in the 0.5 km^2 neighbourhood, frequented by an estimated 200 thousand customers every weekend, most of them tourists (Vendely, quoted in Jovanoski, 2013). Hungary experienced a 7% growth in tourism since 2010 and a 21% growth in nights spent in the country (Budapest Business Journal, 2017). Meanwhile, budget airlines such as Wizz Air have consciously promoted Eastern European locations as destinations offering low priced alcohol for more than a decade (TOB, 2006). This played a major role in the growth of the tourism-related NTE in the party quarter and our research shows that NTE entrepreneurs are primarily targeting tourists now. Prices are increasing and English is often the default language spoken to customers when they enter bars and restaurants.

Neo-patrimonial regulation and the weak public sector

National and local level regulation under neo-patrimonialism

The national level regulation of commercial activities[7] in 2009 was an important factor for the emergence of the party quarter. According to off-the-record sources, this regulation was also the result of non-transparent lobbying of real estate investors. It took away the preliminary licensing right of local authorities, and only the regulation of opening hours remained in their hands. The regulation made it impossible to control the number, size and type of the venues (see Roberts et al., 2006, p. 1118).

In late 2012, a modification of the national level legislation allowed the police to close down venues temporarily or permanently. The legislation was issued by an MP, an important member of the ruling party Fidesz, who was at that time also the Mayor of District V, which is the most central and traditional tourist area of Budapest. As a result, 22 bars received a decree from the police for temporary closure or restriction of opening hours in the party quarter. This caused a strong reaction in the media, and the MP told the press it was not his intention to close venues in the party quarter. Certain entrepreneurs in the party quarter have close relations with the highest level of the ruling party and negotiated with them directly about this issue. In early 2013, the MP issued a modification that allowed longer opening hours for a low amount of extra taxation in 'heritage' inner city areas such as almost the whole party quarter with the exception of a few streets.

In response to the modification of the national level legislation in 2013 (mentioned above) the local authority of District VII allowed a 6am closing time in the party quarter, but in less central parts of district the closing time was set to midnight. The neo-patrimonial element here is the importance of party hierarchy and the supermajority in the parliament that allows the creation of national level legislation for very particular interests. This is a fundamental element of illiberalism: the national level legislation changes for the interests of important cadres of the ruling party (see above Szelényi & Csillag, 2015, p. 24). According to state theories of neoliberalism, it is 'most saliently' experienced at the urban level (Jessop, 2002, p. 452) and decisions are often made at the municipal level (ibid. p. 454). However, the process of regulation described above hardly fits into this narrative. The whole process was directed on the national level, and municipalities could only adapt to the decisions already made above their head. In the illiberal context the sovereignty of local authorities decreased by law and in practice as well (Hegedűs & Péteri, 2015).

The urban and the municipal level policy

Neoliberal and entrepreneurial strategies of tourism promotion were absent at the city level in Budapest (see Smith et al., 2018). In Berlin, for example, the neoliberal city government strongly promotes tourism and the location of bars is more dispersed across districts (Weber & Gallep, 2017). In Budapest, however, because of the missing city level strategy, it was possible to issue restrictive NTE policies in adjacent inner city districts, and consequently, the party quarter became a concentrated problem area. Right-wing mayors of other districts did not follow a pro-enterprise neoliberal agenda at all. 'Recreat[ing] social order' (Bernt, 2012, p. 3046) was much more important to gain the conservative votes in their district.

Typically in adjacent districts to the party quarter, the opening hours of bars are decided discretionally by a committee or new venues are not allowed to stay open until late. Already granted opening hours, however, were not restricted. The case of a cultural and activist centre in the adjacent District VIII serves as a good example of the political decisions driving regulations and their enforcement. The centre hosts a number of NGOs that are considered the 'enemy' in everyday illiberal political propaganda. Their venue generates income from hospitality and parties. While other venues also cause considerable noise, only this centre is threatened with closure repeatedly by the local authority using deliberate misinterpretation of the regulations[8] and even buying out the building from the current owner was proposed by the local authority.[9] The attitude of the local authority is well illustrated by the peculiar concentration of street patrols: '[t]his summer every night a car of the municipal street wardens was parking before our door together with a policeman ... at other venues ... there is no such constant presence.'

The rules change quickly and unpredictably outside the party quarter. Investors have been preparing a large disco in District VIII for years now, and negotiated with the local authority several times about the venue. In late 2018, however, the local authority suddenly decided to apply the 12pm closing time of bars and shops to dance music venues as well. Until then they had been exceptions to this closing time regulation. 'That's about being in good relations with the local authority' – commented a bar owner. These elements of the local policy do not fit the entrepreneurial urban governance framework. At the same time, particular treatment of certain bars shows the importance of politically and ideologically motivated decisions under an illiberal regime.

Lack of state capacity

Neo-patrimonial governance does not mean a very effective 'penal state'. Police are notoriously understaffed for the task even after proportionally major improvements. When the district police chief admitted the lack of resources for the party quarter semi-publically in 2017, he was dismissed from his position immediately. Consequently, drinking on the streets, prostitution and open drug dealing is common in the party quarter. On the other hand, according to the results presented by the police, crime rates are declining in the party quarter[10], a 'fact' the residents strongly dispute. In a public hearing in late 2018, the office of the legal department of the local authority admitted that there are no statistics about noise complaints against bars in the party quarter (repeated complaints could lead to the closure of the venues). So far, only 8 bars have been prosecuted for noise complaints from the hundreds located there, and these legal processes can last for years. Residents have since protested in public hearings for the resignation of the leader of the legal department and accused him of corruption.

Interest articulation of stakeholders in a neo-patrimonial context

Residents

After further loosening the regulation of the NTE in 2013, the ruling party lost many seats in the local council in 2014, but the Fidesz member mayor remained in power. Only in 2017 was a neighbourhood level movement for a 12pm closing time organised with politically experienced leaders (for the similar case of Prague see Pixová & Sládek, 2017, p. 82). Most supporters of the Élhető Erzsébetvárosért (Liveable Elisabeth-town) movement are longer term residents in the area with various social backgrounds. They were able to pressure the local authority with the help of the press and demonstrations. Instead of complying with their request, the local authority decided to hold a district level referendum about the 12pm closing time. It meant that a large number of the District's residents who were not involved in the problem could vote on it. A political campaign by the ruling as well as the opposition parties was completely absent. The turnout was very low, which meant the referendum was invalid, and regulations have not changed since. After this defeat, the Liveable Elisabeth-town movement split in two. There were disputes within the group about whether to campaign for a 12pm closing time during the referendum under these circumstances, and there were also left-right political tensions between members. The Liveable Elisabeth-town movement was tricked into an already lost battle and it broke their morale.

Entrepreneurs

In early 2010, the entrepreneurs started to organise a civil group called 'Azért' to lobby against the early closing time already introduced in adjacent District VI. When the national regulation became stricter in 2012 and a press controversy arose, the local authority officials and Fidesz politicians negotiated with the Azért, however, there were only a few more than 20 bars involved in the group from the hundreds in the area. The lawyers of the organisation played a major role in the codifications process of the national law introduced in 2013 (see above). On their suggestion, the extra tax was limited by a cap and the local authority was not even aware of this amendment.[11] The local regulation divided the opening hours geographically which also meant that some members of the Azért (outside the party quarter) had to close at midnight. The lobbying of bars was divided during the whole process and trust between entrepreneurs was very low. This is also visible from the other side of the barricade: 'It is obvious that the local authority does not serve the interests of the entrepreneurs or not all of them' – said one of the residents fighting against the party quarter. Similarly, in other districts, certain entrepreneurs could cooperate well with the district mayors, while others could not follow the proceedings as closely. For example, the above mentioned activist centre remained an outsider: 'we also heard later that it came up [in the meetings] not to mention our name in this association [that was lobbying against the restrictive local regulation]'. The unified lobbying of the entrepreneurs stands far from reality. The formally civil initiative of the Night Mayor of Budapest (NMB) is backed by the largest and most successful bars. They suggest the dispersion of the NTE to the whole inner city and higher standards for bars. Admittedly, this would mean the closure of many cheaper venues primarily serving alcohol, so their suggestion would also mean narrowing the competition. The moderate political impact of the NMB is well illustrated by the behind the scenes comment of the right-wing mayor of Budapest: 'as far as I know, I am the only mayor of this city.'

Political actors

The weak opposition in this illiberal democracy have difficulties in positioning themselves in the debate. Liberal parties, for example, were afraid to lose votes at the national level if they fully backed the anti-NTE movement. The anti-political sentiments of the residents and the political divide within the residential movement Liveable Elisabeth-town movement inhibited strong cooperation between residents and the opposition. 'I experienced that these representatives either don't want to help or [were] incapable' – commented an active resident. After the divide of the Liveable Elisabeth-town, the part of the movement under the original name now plans to become an official civil organisation and gain seats in the municipal council. These activists were typically less openly critical of the Fidesz government and framed the conflict in the party quarter as a local issue, separated from traditional left-right or regime-opposition relations. They are certainly not against entrepreneurs in general, and campaign for high status venues such as expensive restaurants instead of cheap bars, as these would supposedly cause less noise. To call this residential movement resistance against neoliberalism would be quite far-fetched. On the other hand, they are also not high status new residents trying to displace the bohemian atmosphere of the district. They are rather long term residents from various social backgrounds trying to resist the displacement pressures of NTE related tourism.

Exclusion of residents

The neo-patrimonial privatisation of housing in the 1990s resulted in low exchange value of apartments in dilapidated inner city buildings. However, after the tourism boom, these apartments could guarantee substantial yields after turning them into STRs and hostels. The private rental market is insufficiently regulated (Csizmady et al., 2017), and many of the long term rentals and STRs exist on the underground market. Tax authorities first threatened STR entrepreneurs with huge fines, but tax control also lacks resources. Eventually, the taxation of STRs simplified and decreased[12] in the hope of making the legal operation more desirable for entrepreneurs. Together with NTE related tourism, STR displaces long term rentals and owner occupiers, even in expensive newly built or renovated inner city buildings (also adjacent to the party quarter). Speculative price increases in the inner city of Budapest raised prices rapidly year by year and inner city housing is not a realistic option for most Hungarians anymore. The lively and socially mixed (Kovács et al., 2013, p. 1.) inner city fabric is slowly disappearing.

Conclusions

Tourism and NTE policies are currently virtually non-existent in Budapest at the city level (Smith et al., 2018). Decisive 2009 national level legislation of opening hours in the inner city was codified without local political control and was modified according to personal interests of the ruling party in 2012 then favoured entrepreneurs in 2013. Restrictive measures against the NTE at the municipal level in adjacent districts to the party quarter are motivated by the goal of keeping 'social order' (Bernt, 2012, p. 3046) and appeal to the conservative voters. Certain entrepreneurs could maintain their position, but their lobbying activity is also far from unified. Rather, personal, patrimonial relations are at play. Meanwhile self-regulation and self-governance of entrepreneurs is not supported politically. The importance of national level regulation goes against the dominant narrative of neoliberal governmentality, preferring local self-governance (Jessop, 2002, p. 454). This makes local interest articulation of residents strongly limited, while democratic institutions such as the referendum are rendered ineffective because of the political passivity of opposition parties and residents.

Phenomena such as property price increases, conflicts over noise, and the decreasing quality of life and exclusion of long term residents from the inner city are in a sense similar to cities with conscious (neoliberal) political support of the NTE and tourism (such as Berlin, Novy, 2017; or Barcelona Nofre, Giordano, et al., 2017). What differentiates the neoliberal examples from our case is the chance of a democratic debate between stakeholders and political representatives over an existing planning or political agenda – at least, in principle. The anti-STR measures taken in Berlin and Barcelona (and other cities) are maybe insufficient to stop touristification but the question is certainly part of the political discourse and electoral campaigns. In our context, the dismantling of liberal democracy strongly limits legal contestation and acquiring political support for certain interests. Our case study sheds light on the importance of planning and democratic political control of urban issues. Despite escalating public tensions in Budapest, gentrification and tourism march relentlessly on without any apparent measures to control or contain them, while the NTE and overtourism are concentrated in the former Jewish ghetto of Budapest.

Notes

1. Short-term rentals
2. https://www.citylab.com/life/2018/03/berlin-airbnb-vacation-rental-regulation-law/556397/
3. https://www.citylab.com/life/2018/06/barcelona-finds-a-way-to-control-its-airbnb-market/562187/
4. Many thanks for this insight to Jordi Nofre.
5. The power of the city of Budapest is limited to the infrastructural network and certain architectural standards, but district authorities are in charge of every other issue within their jurisdiction.
6. https://index.hu/belfold/2013/10/10/hunvald_masodfokon_is_masfel_evet_kapott/
7. 210/2009. (IX. 29.) Government regulation of the 2005. CLXIV. Law about commercial activities
8. https://www.helsinki.hu/torvenyserto-volt-az-aurora-kioszk-bezaratasa/
9. https://magyarnarancs.hu/lokal/most-megvenne-az-onkormenyzat-az-aurora-epuletet-csak hogy-az-bezarjon-113798
10. https://444.hu/2018/04/18/a-hatosagok-nem-talalaljak-a-problemas-bulizokat-a-bulinegyedben
11. https://index.hu/belfold/2014/01/09/elszamoltak_a_rogan-fele_romkocsmaadot/
12. https://www.nav.gov.hu/nav/ado/afa080101_hatalyos/Online_szallashely_szolgaltatas.html

Disclosure statement

No potential conflict of interest was reported by the authors.

ORCID

Gergely Olt http://orcid.org/0000-0001-9816-5601

References

Barnett, C. (2005). 'The consolations of "neoliberalism". *Geoforum; Journal of Physical, Human, and Regional Geosciences*, *36*, 7–12.

Barnett, C. (2010). Publics and markets: What's wrong with neoliberalism? In S. J. Smith, R. Pain, S. A. Marston, & J. P. Jones III (Eds.), *The Sage handbook of social geography* (pp. 269–296). London: Sage.

Bernt, M. (2012). The 'double movements' of neighbourhood change: Gentrification and public policy in Harlem and Prenzlauer Berg. *Urban Studies*, *49*, 3045–3062.

Bernt, M. (2016a). Very particular, or rather universal? Gentrification through the lenses of Ghertner and lópez-Morales. *City*, *20*(4), 637–644.

Bernt, M. (2016b). How post-socialist is gentrification? *Observations in East Berlin and Saint Petersburg. Eurasian Geography and Economics*, *57*(4–5), 565–587. DOI:10.1080/15387216.2016.1259079

Bernt, M., Gentile, M., & Marcinczak, S. (2015). Gentrification in post-communist countries: An introduction. *Geografie*, *120*(2), 104–112.

Budapest Business Journal. (2017). Hungary's tourism continues to pick up in 2016, February 10th, Retrieved from https://bbj.hu/economy/hungarys-tourism-continues-to-pick-up-in-2016_128614

Bürkner, H., & Totelecan, S. G. (2018). Assemblages of urban leisure culture in inner-city Bucharest. *International Journal of Urban and Regional Research*, *42*, 786–806. doi:10.1111/1468-2427.12630

Collier, S. J. (2012). Neoliberalism as big Leviathan, or … ? A response to Wacquant and Hilgers. *Social Anthropology*, *20*, 186–195.

Cócola Gant, A. (2016). Holiday rentals: The New gentrification Battlefront. *Sociological Research Online*, *21*(3), Retrieved from http://www.socresonline.org.uk/21/3/10.html

Csanádi, G., Csizmady, A., & Olt, G. (2012). *Átváltozóban [transformation]* (p. 268). Budapest: Eötvös Kiadó.

Csizmady, A., Hegedűs, J., & Nagy, G. (2017). The effect of GFC on tenure choice in a post-socialist country – the case of Hungary. *International Journal of Housing Policy, 17*(2), 249–275.

Czirfusz, M., Horváth, V., Jelinek, C., Pósfai, Z., & Szabó, L. (2015). Gentrification and Rescaling urban governance in Budapest-józsefváros. *Intersections, EEJSP, 1*(4), 55–77.

Evans, G. (2017). Tourism and the Night – Hold back the Night Presentation at *Tourism And The Night An Interdisciplinary Symposium* 14th and 15th July, University Of Westminster.

Ferguson, J. (2010). The Uses of neoliberalism. *Antipode, 41*, 166–184.

FHB. (2016). FHB lakásárindex – lakáspiaci kilátások [Housing price index – housing market prospects]. Retrieved from http://www.fhbindex.hu/FHB-Index

Földes, A. (2003). 'Láthatatlan' alternatív szórakozónegyed, belső udvarokban [Invisible alternative party quarter in the inner courtyards], *Index*, 10 June Retrived from http://index.hu/politika/bulvar/alter0608/;

Gotham, K. F. (2005). Tourism gentrification: The case of New Orleans' Vieux Carré (French quarter). *Urban Studies, 42*(7), 1099–1121.

Gravari-Barbas, M., & Guinand, S. (2017). *Tourism and gentrification in Contemporary Metropolises*. London: Routledge.

Hadfield, P., Lister, S., & Traynor, P. (2009). This town's a different town today: Policing and regulating the night-time. *Criminology and Criminal Justice, 9*, 465–486.

Hae, L. (2011). Dilemmas of the nightlife fix. Post-industrialization and the gentrification of nightlife in New York city. *Urban Studies, 48*, 3444–3460.

Hamnett, C. (2003). *Unequal city: London in the Global Arena*. London: Routledge.

Harvey, D. (2005). *A brief history of neoliberalism*. Oxford: Oxford University Press.

Hegedűs, J., & Péteri, G. (2015). Közszolgáltatási reformok és a helyi önkormányzatiság [Reforms of public services and local autonomy]. *Szociológiai Szemle [Sociological Review], 25*(2), 90–119.

Hegedűs, J., & Tosics, I. (1998). A közép-kelet-európai lakásrendszerek átalakulása [The transformation of the Central-European Housing System]. *Szociológia [Sociology]*.

Hubbard, P. (2017). *The battle for the high street retail gentrification class and Disgust*. London: Palgrave Macmillan.

Islam, T. (2005). Outside the core gentrification in Istanbul. In R. Atkinson & G. Bridge (Eds.), *Gentrification in a global context: The new urban colonialism* (pp. 123–138). London: Routledge.

Jávor, I., & Jancsics, D. (2016). The role of power in organizational corruption: An empirical study. *Administration and Society, 48*(5), 527–558.

Jessop, B. (2002). Liberalism, neoliberalism, and urban governance: A state theoretical perspective. *Antipode, 34*, 452–472.

Jovanoski, K. (2013). Budapest ruin pubs offer peep into the past, Al Jazeera, 16 April, Retrieved from https://www.aljazeera.com/indepth/features/2013/04/201341583847649333.html

Kębłowski, W., & Van Criekingen, M. (2014). How 'alternative' alternative urban policies really are? *Métropoles*, (15). Retrieved from http://metropoles.revues.org/4994

Kovacs, Z., Wiessner, R., & Zischner, R. (2015). Beyond gentrification: Diversified neighbourhood upgrading in the inner city of Budapest. *Geografie, 120*(2), 250–273.

Kovács, Z., Wiessner, R., & Zischner, R. (2013). Urban Renewal in the inner city of Budapest: Gentrification from a Postsocialist perspective. *Urban Studies, 50*(1), 22–38.

Lugosi, P., Bell, D., & Lugosi, K. (2010). Hospitality, Culture and Regeneration: Urban Decay, Entrepreneurship and the 'ruin' bars of Budapest. *Urban Studies, 47*(14), 3079–3101.

Nofre, J., Giordano, E., Eldridge, A., Martins, J. C., & Sequera, J. (2017). Tourism, nightlife and planning: Challenges and opportunities for community liveability in la Barceloneta. *Tourism Geographies*, DOI: 10.1080/14616688.2017.1375972

Nofre, J., Sanchez-Fuarros, I., Carlos Martins, J., Pereira, P., Soares, I., Malet-Calvo, D., … Lopez Diaz, A. (2017). Exploring Nightlife and urban change in Bairro Alto,Lisbon. *City & Community, 16*(3), 330–344.

Novy, J. (2017). The selling (out) of Berlin and the de- and re-politicization of urban tourism in Europe's 'capital of Cool'. In C. Colomb, & J. Novy (Eds.), *Protest and resistance in the tourist city* (pp. 52–72). London: Routledge.

Ong, A. (2007). Neoliberalism as a mobile technology. *Transactions of the Institute of British Geographers, 32,* 3–8. doi:10.1111/j.1475-5661.2007.00234.x

Pinkster, F. M., & Boterman, W. R. (2017). When the spell is broken: Gentrification, urban tourism and privileged discontent in the Amsterdam canal district. *Cultural Geographies, 24*(3), 457–472.

Pixová, M., & Sládek, J. (2017). Touristification and awakening civil society in post-socialist Prague. In C. Colomb & J. Novy (Eds.), *Protest and resistance in the tourist city* (pp. 73–89). London: Routledge.

Roberts, M. (2009). Planning, urban design and the night-time city: Still at the margins? *Criminology and Criminal Justice, 9,* 487–507.

Roberts, M., Turner, C., Greenfield, S., & Osborn, G. (2006). A Continental Ambience? Lessons in Managing alcohol-related Evening and night-time Entertainment from Four European Capitals. *Urban Studies, 43*(7), 1105–1125.

Román, A. (1998). Madách Imre, avagy egy sugárút tragédiája" [Imre Madách or a tragedy of a promenade] Budapesti Negyed 18–19, Retrieved from http://epa.niif.hu/00000/00003/00015/roman.htm

Sipos, A., & Zolnay, J. (2009). Ingatlanpanama a régi pesti zsidónegyed elpusztítása [Real-estate corruption the destruction of the old Jewish Quarter] *Beszélő* 14/5, Retrieved from http://beszelo.c3.hu/cikkek/ingatlanpanama

Slater, T. (2017). Clarifying Neil Smith's rent Gap theory of gentrification. *Tracce Urbane Italian Journal of Urban Studies, 1*(Giugno), 83–101.

Smet, K. (2016). Housing prices in urban areas. *Progress in Human Geography, 40*(4), 495–510.

Smith, A., & Pappalepore, I. (2015). Exploring attitudes to edgy urban destinations: The case of Deptford, London. *Journal of Tourism and Cultural Change, 13*(2), 97–114.

Smith, M. K., Egedy, T., Csizmady, A., Jancsik, A., Olt, G., & Michalkó, G. (2018). Non-planning and tourism consumption in Budapest's inner city. *Tourism Geographies, 20*(3), 524–548.

Smith, N. (1979). A theory of gentrification: A back to the city movement by capital not people. *Journal of the American Planning Association, 45*(4), 538–548.

Smith, N. (1982). Gentrification and Uneven development. *Economic Geography, 58*(2), 139–155.

Smith, N. (1996). *The new urban frontier: Gentrification and the Revanchist city.* New York, NY: Routledge.

Stark, D., & Vedres, B. (2012). Political holes in the economy: The business network of partisan firms in Hungary. *American Sociological Review, 77*(5), 700–722.

Sykora, L. (1993). City in Transition: The role of rent gaps in Prague's Revitalization. *Tijdschrift Voor Economische en Sociale Geografie, 84*(4), 281–293.

Sykora, L. (2005). Gentrification in post-communist cities. In R. Atkinson & G. Bridge (Eds.), *Gentrification in a Global context The new urban colonialism* (pp. 91–106). London, New York: Routledge.

Szelényi, I. (2015). Capitalisms after Communism. *New Left Review, 96,* 39–51.

Szelényi, I. (2016). Weber's theory of domination and post-communist capitalisms. *Theory and Society, 45*(1), 1–24.

Szelényi, I., & Csillag, T. (2015). Drifting from liberal democracy. Neo-conservative ideology of managed illiberal democratic capitalism in post-communist Europe. *Intersections. East European Journal of Society and Politics, 1*(1), DOI: https://doi.org/10.17356/ieejsp.v1i1.28

Tourist Office of Budapest (TOB). (2006). *10 years Anniversary of the tourism office of Budapest.* Budapest: TOB.

Tuvikene, T. (2016). Strategies for comparative urbanism: Post-socialism as a de-territorialized concept. *International Journal Of Urban And Regional Research, 40,* 132–146. doi:10.1111/1468-2427.12333

Wachsmuth, D., & Weisler, A. (2017). Airbnb and the Rent Gap: Gentrification Through the Sharing Economy., Retrieved from https://davidwachsmuth.files.wordpress.com/2017/07/wachsmuth-weisler-airbnb-draft.pdf

Weber, C., & Gallep, P. (2017). Berlin nights – showcase for understanding night time dynamics, conflicts and concepts, Presentation at *Tourism And The Night An Interdisciplinary Symposium* 14th and 15th July, University Of Westminster.

Commuting and the urban night: nocturnal mobilities in tourism and hospitality work

Anna Plyushteva

ABSTRACT

The importance of the tourism and hospitality sectors to the urban night and the economies of cities has been well-documented. However, commercialised night-time leisure as a timespace of work remains relatively underexplored. A greater understanding of bars, hotels, restaurants and clubs as workplaces as well as leisure destinations can contribute to conceptualising the potential of urban nights to be more diverse and inclusive. This paper examines the challenges faced by tourism and hospitality workers who need to commute at night. Drawing on a mixed-method study of Sofia, Bulgaria, and Brussels, Belgium, the paper explores the barriers encountered and resources mobilised to make the nocturnal commute possible. The dimensions discussed include the need to continuously adjust one's commute; the economic and non-economic costs of night commuting; cycling at night; and the impact of collegial and employer relations on negotiating the night commute. The paper demonstrates that while tourism and hospitality workers mobilise a range of economic, organisational, affective and physical resources in order to make night commutes possible, much remains to be done by urban authorities and transport planners to ensure inclusive nocturnal mobilities which attend to work as well as leisure journeys.

Introduction

The urban night is often thought of as a timespace of leisure and freedom, an opportunity to escape the pressures of work and the routines of everyday life. This notion is reflected in the image many European cities have been keen to commercialise in their 'night-time economies,' with authorities and tourism and hospitality businesses positioning cities after dark as destinations of entertainment, indulgence and escape (see, among others, Hollands & Chatterton, 2003; Roberts & Eldridge, 2007; Shaw, 2018). However, the work required to sustain night-time destinations such as restaurants, bars and hotels, and the conditions in which nightlife employment is performed, have received less attention (cf. Liempt & Aalst, 2015; Monaghan, 2002). The present paper contributes to a body of work which explores the particular and often problematic demands which commercialised night-time leisure places on the physical and emotional labour of hospitality workers (see Guerrier & Adib, 2003; McDowell, Batnitzky, & Dyer, 2007). Clearly, a great diversity of occupations and working conditions exists in tourism and hospitality (Baum, 2007). Nevertheless, work in these sectors remains frequently associated with long and unpredictable shift hours, low and/or precarious pay, and frequent change of workplace locations (Vladimirov & Vedar, 2016). In this context, the journey to work can present particular economic, physical and emotional demands for tourism and hospitality employees working late hours.

This paper discusses the night commute as one aspect of tourism and hospitality work in night-time cities. Drawing on a mixed-method study of Sofia, Bulgaria and Brussels, Belgium, the paper aims to explore the challenges faced by night-time commuters and the individual and shared resources which they mobilise to make their nocturnal journeys possible.

As mobilities research has repeatedly demonstrated, commuting is not simply a goal-driven movement from A to B, but a particular mode of inhabiting the city, replete with interactions, negotiations, frictions and emotions (Bissell, 2018; Middleton, 2009). The paper aims to enrich the night-time economy literature by attending to the night commutes of tourism and hospitality workers as a significant aspect of their labour, and thus of night-time urban life. The paper draws on theoretical approaches to the subjective and situated experiences of commuters developed by mobilities scholars (Corvellec & O'Dell, 2012; Plyushteva, 2016; Waitt, Harada, & Duffy, 2017). In the past two decades, a number of studies in tourism and hospitality have explored issues around mobility relevant to these sectors. Topics have included the journeys of tourists, the mobilities of trends and ideas, as well as the role of international migration in shaping the tourism and hospitality workforces in many locations (Duncan, Scott, & Baum, 2013; Edensor, 2007; Hall, 2005). However, the seemingly mundane mobilities involved in commuting to tourism and hospitality jobs have received limited attention, reflecting also a broader scarcity in research on hospitality workplaces, working conditions and everyday work practices (Baum, Kralj, Robinson, & Solnet, 2016; Ioannides & Zampoukos, 2018; Jordhus-Lier & Underthun, 2015).

A mobilities framework draws attention to the perspective of workers of late shifts, for whom the urban night is not a spatio-temporal container separated from the mundane pressures of the daytime. At the same time, while the day and night may be indivisible in terms of human experience, urban authorities and transport planners take very different approaches to them in terms of transport provision. To address this, the paper makes use of insights from studies of transport exclusion (Lucas, 2012; Schwanen et al., 2015), in order to make visible the specific barriers which night commuters encounter. While transport exclusion may manifest itself along many different dimensions in both daytime and night-time cities, the urban night presents particular barriers to mobility. Public transport is scarce or absent; taxis are more expensive; people on the move and women in particular are more likely to feel unsafe and vulnerable (Collectiu Punt 6, 2017). Tourism and hospitality workers doing late shifts perceive and tackle these issues in ways shaped by the particularities of their workplace, such as uncertain working hours or expectations regarding personal presentation. As night mobilities remain inaccessible and precarious for many, greater attention to the needs of those who are mobile at night for work rather than leisure is needed if nocturnal cities are to become more inclusive (Hadfield, 2015).

Research locations and methods

The research presented in this paper is part of a wider comparative project which examined shift work, commuting and night-time public transport provision in Brussels and Sofia. The capital cities of Belgium and Bulgaria each have populations of approximately 1.3 million inhabitants (National Statistics Institute, 2018; StatBel, 2018). Hospitality and tourism work was selected as a case study for this project due to the frequent combination of around-the-clock shift work, uncertain hours and low and/or irregular incomes which workers in these sectors experience. The study focussed on small and medium-sized independent businesses in these sectors, covering hotels, restaurants, bars, cafes and night clubs. Tourism has greater importance to the Brussels economy, with 4.3 nights per resident spent in tourist accommodation in 2016 compared to Sofia's 1.4 (Eurostat, 2018). However, tourism and hospitality have greater relative importance for employment in Sofia, with an estimated one-third of jobs found in the trade, transport, hotels and restaurants sectors in 2011 (Eurostat, 2018).

The hours between 23:00 and 05:00 were taken as the basis for defining the night commute, as at the start of the study, public transport services were unavailable on weeknights during those hours in either Sofia or Brussels.[1] In Brussels, a night bus service has been operating on Friday and Saturday nights since 2007, with last buses departing the city centre around 2.45am. In addition, a shared night taxi service called Collecto[2] operates all nights of the week. In Sofia, an all-week night bus service was introduced in April 2018 in response to public pressure (Zografsky, 2018). As data collection for this paper took place between May 2017 and May 2018, its introduction had not yet impacted on respondents' commuting practices. In future, its longer operating hours compared to the Brussels service is likely to influence night mobility to a greater extent, although both services are limited in their spatial reach. In addition, both night bus services require the purchase of special tickets, which are not integrated in daytime public transport travelcards, and are also more expensive. Both city authorities have banned the operation of ride-hailing app Uber, although the company's premium service continues to operate in Brussels (with limited utility for tourism and hospitality night commuters due to its price). Brussels also offers a city bike-sharing service which is available 24/7, whereas no such scheme has been introduced in Sofia as of August 2018.

The study relied on a combination of quantitative and qualitative methods. Initially, online survey questionnaires were distributed to tourism and hospitality employers and employees using a snowball sample, ensuring that responses broadly reflected the spatial distribution of hospitality and tourism business across the two cities (in both cases, venues are concentrated in the city centre, but prominent secondary centres of night leisure can be found in peripheral neighbourhoods). The survey covered questions about commuting routes, modes, costs and durations; barriers faced; employer policies and facilities relevant to commuting; and demographic data. 49 employer/manager and 92 employee questionnaires were completed. Follow-up in-depth interviews were carried out with 20 respondents (11 employees and nine owners/managers), with interviews lasting between 40 and 90 min and allowing participants to discuss in greater depth their survey responses. Interview respondents were rewarded with a shopping voucher for their time. Data were analysed using data analysis software and are reported here in anonymised form using pseudonyms.

Results

The continuous re-negotiation of night commutes

The commute is often thought of as a predictable and stable everyday practice (Anable, 2005; Schlich & Axhausen, 2003). Researchers have often conceptualised the commute as an arrangement based on an initial one-off rational evaluation of costs and benefits, which is then repeatedly performed in a routine, automatic manner (cf. Bissell, 2011, 2018; O'Dell, 2009). However, the journeys of the tourism and hospitality workers interviewed for this study defy this characterisation. For many participants, the night-time journey was central to making decisions on whether to start a particular job in tourism and hospitality, and then required continuous adaptation and renegotiation.

When Georgiana started working as a waitress during her university studies, she specifically chose a restaurant which was within a short walk from her parents' house:

> It took five minutes to walk. I wanted it to be close, because I knew I will have to work night shifts. (Georgiana, Brussels)

The need to travel at night can limit what positions a worker perceives as accessible to them. However, the majority of participants reported that the issue of the night commute was not raised by either applicant or employer at the recruitment stage, resulting in the need to continuously try out possible arrangements, and negotiate the night commute as part of gradually settling into a workplace. This involved learning about transport options, gauging the availability of support from the employer and the possibilities for sharing lifts with colleagues, and importantly, the level of earnings which could be expected – and which over the first few weeks determined whether certain options such as a taxi at night would be affordable, or not. Georgiana described this dynamic in the context of one of her subsequent appointments in the sector:

> - When you applied for the job, did you discuss night transport with the owner or manager?
>
> - I didn't ask, and they didn't say anything. It's like it wasn't an issue. That's interesting. It should be discussed. It was a bar, I finished at 4am. Usually, I walked to my house. It was a 40-minute walk. That's why I stopped, because it was too much. I didn't want to take a taxi. It made no sense financially – a taxi would cost more than 10 euros, that's more than an hour's work. It turned out it was too much. I worked there for three weeks. (Georgiana, waitress, Brussels)

For many research participants, working evening and night shifts meant that the journey to work, and the return journey home, differed significantly. Research participants who worked late hours typically found that the two parts of their commute required their own separate arrangements in terms of transport mode, duration, route and payment. While the majority of survey respondents found daytime journeys to be relatively straightforward, the night-time journey required continuous negotiation:

> Occasionally, the manager gives people a lift [at night], and since I live on her route, she drops me off, which is great for me. However, this is not anything that is organised systematically: if it happens, great, if not, I have to fend for myself. No money is given for the taxi fare home. Sometimes I share a taxi with some colleagues. But it all depends which shift you are in, which people you end up in a shift with. (Julia, waitress, Sofia)

On some occasions, both Georgina and Julia had experienced needing to walk home from work at night through areas which made them felt unsafe or uncomfortable. This reflects the concern raised by some mobilities and transport researchers that enduring assumptions about the commute often obscure important gendered dynamics (England, 1996; Schwanen, 2007). In the context of daytime commuting, the portrayal of the outward and inward part of the commute as mirror images of each other can make invisible the fact that women in particular are likely to make complex trips (also known as trip-chaining) as part of returning home from work, collecting children from school, shopping, or undertaking other unpaid care work (Gonzales & Daganzo, 2013; Kwan, 1999). Similarly, the gendered mobilities of night workers are excluded from straightforward conceptualisation of everyday mobilities as A to B and B to A. By contrast, the interviewed women – and on two occasions, men – described feelings ranging from unease to fear in relation to walking home at night. Thus, the need to repeatedly reorganise and adjust mobility arrangements during the night hours could represent a significant burden, in terms of planning, worrying, dealing with uncertainty, and additional expenses.

The costs and affects of public transport at night

One key policy measure which has the potential to improve the ability of tourism and hospitality workers to access and sustain employment is introducing night-time public transport services, or ensuring that night work is reflected in the routes and timetables of existing night services. However, the mere availability of such services is not a sufficient condition in itself, as respondents frequently pointed out the importance of affordability. With 35% of surveyed employees relying on taxis or private cars to return home, night-time public transport had the potential to reduce a significant financial burden for night commuters. However, the specifics of pricing and scheduling were critical to whether the service was seen as accessible and beneficial, or not:

> It took me a long time even to figure out how it works. I wasn't sure if I needed to buy a separate ticket. And then, I feel like the last Noctis[3] is earlier and earlier all the time. Once, at half past two I already couldn't take it. It's a long time until the first regular bus in the morning. Plus, when I have weekday shifts, it doesn't work for me. (Thomas, hotel receptionist, Brussels)

In Brussels, only one in five survey respondents reported making use of the city's limited night bus service at least occasionally. At a fixed price of 5 euro per ride, the Brussels Collecto shared night taxi service was described by interviewees as suitable for people on a night out but too expensive for regular commutes. Similarly to Collecto, Sofia's new night bus service operates a fixed price, which at approximately one euro is 25% higher than that of a daytime public transport ticket. With existing daytime travelcards not accepted, night bus commuters therefore shoulder two sets of expenses if they also buy daytime public transport tickets or travelcards.

Incorporating night-time public transport tickets into the existing travelcards, rather than charging them as a separate service, thus appears to be an essential first step in reducing the challenges faced by nocturnal commuters. However, for those working in the tourism and hospitality sectors, this may involve additional barriers and risks. While many 9-to-5 commuters can make substantial savings by purchasing a monthly or annual travelcard, this is predicated on mobility needs which remain fairly stable and predictable over time. This was not the case for 65% of survey respondents:

> At the moment, I work as a waitress in [Restaurant], I also take occasional shifts as a washer, and I have another job, in a pub in the city centre. But things change all the time. For me, it's easier not to buy a travelcard, because I don't always use public transport that much. To be honest, I don't even buy a ticket. I got a fine once, but it still makes more sense for me financially to pay the occasional fine. (Tanya, waitress, Sofia)

Thus, unpredictable shifts, changing workloads, incomes and employer demands impacted on the ability to plan ahead for transport costs. In both Sofia and Brussels, many restaurants and bars close for a period over the summer months – in some cases, as long as 10 weeks. In the month of January, operations can be scaled back, with many employees working a reduced number of shifts. As a result, an investment in an annual travelcard could be too risky given the large initial expense involved.

Access to student travelcards in both Brussels and Sofia could provide substantial savings in both cases, and highlighted the on-going negotiations commuting entailed. With 32% of the survey respondents combining tourism and hospitality work with study, a student travelcard was a key resource for mobility, even if not useable for night travel. For several participants, it made the difference between an acceptable income from a job in these sectors, and earnings which would be too low to justify the cost of commuting. However, once working students graduated and became ineligible for a reduced-price travelcard, making new commuting arrangements could result in significant upheaval:

> The advantage, when you are a student, is that for three years the travelcard costs you 50 euro per year. I was already working in the hotel as a student and I am still there, but now, the monthly travelcard costs me 50 euro per month. This is quite a difficulty when you finish your studies. Ok, you are going to earn a salary, but it is not easy, it's a difficult financial transition. It's a huge change. 50 per year versus 50 per month. (Thomas, hotel receptionist, Brussels)

Although economic costs were a principal concern for many respondents, using public transport at night is also associated with particular emotional costs which interviewees highlighted. Over and above a financial burden, these journeys often came with sensations and socialities which were not always enjoyable for workers returning home. Following Bissell (2010, 2018), these experiences can be described as the affective atmospheres of public transport, often configured differently and intensified after dark. Such atmospheres could be challenging in ways which are not present for daytime commuters. As explained by Daniel, a bartender in Brussels, sharing a bus or metro train with passengers on a night out could be a tense experience when returning home from a long shift:

> I felt a certain tension between me and other people. In the evenings, especially on weekends. Drunken things. It can be loud – arguments, fights. There were people who have no respect for other people. After a night shift, you just sit there thinking any moment something can happen.

These comments can be interpreted in the context of a wider debate on the ways in which alcohol consumption can shape the experiences of the night-time city (Eldridge & Roberts, 2008). For Daniel, these experiences on the metro ride home contributed to the psychological fatigue he had accumulated in his customer-facing role at the bar. He eventually abandoned public transport, choosing to drive or cycle to work instead. The next section focuses on the arrangements which made a bicycle commute possible when working the late shift.

Complicating the active and sustainable commute with the night-time

In recent years, transport planners in many European cities have sought to develop approaches to commuting which focus on two related agendas: sustainability and active transport. Sustainable commutes are encouraged for their lesser environmental impacts in terms of air pollution, noise and demands on urban space, and have generally focussed on promoting alternatives to private car transport (Banister, 2008; Schwanen, Banister, & Anable, 2011). Relatedly, active commuting by modes such as walking and cycling has attracted policy attention for their health benefits in light of increasingly sedentary urban lifestyles (Gerike et al., 2016). However, both agendas remain largely focussed on the 9-to-5 office worker, and interventions have mostly targeted large white-collar organisations (Van Malderen et al., 2012).

At first glance, walking and cycling can seem particularly well-suited to the night commute, when public transport provision is limited or non-existent. In both Brussels and Sofia, this assumption could be compounded by the fact that many tourism and hospitality workers choose employment located within relatively short distances from their homes (Brandeleer et al., 2016; see also Section 3.1). However, this view must be qualified in a number of ways.

Importantly, the small and medium-sized employers who predominate in this sector are often unable to provide shower facilities and secure bicycle storage, due to lack of space. Only 17% of the employers surveyed provided both of these at their establishments. Combined with the exacting requirements for personal presentation for customer-facing roles, this limitation presented a barrier to commuting by bicycle for several research participants:

> I do it, but most of my colleagues are worried about sweating. Heading into work on my bike, the workout of the muscles, spinning your head this way and that to check for cars, the sweating, the stress, the anxieties of cycling in a road. With all of this, you are on your way to work. You have to be fresh, to be presentable. You are working with customers. I had work clothes to change into, but still. (Daniel, bartender, Brussels)

At the same time, cycling did offer specific benefits for tourism and hospitality workers needing to travel at night. Eleven survey respondents (10% of total) chose cycling as either a main or occasional mode of travelling to work. In the four follow-up interviews carried out among this group, all participants pointed out the affordability of bicycle commuting, compared to taxi and private car, and two of the participants also felt it offered greater personal safety at night compared to walking (see Section 3.1). Interestingly, physical exhaustion was not perceived as a barrier: both cyclists and non-cyclists felt that the physical demands of their work would not prevent them from cycling to work:

> The bicycle is great, it's very healthy. Even though I work on my feet, cycling can actually be really good for tired feet, for the circulation. (Kyle, sous-chef, Sofia)

This is a potentially surprising finding, since efforts to promote active commuting have generally excluded those in physically demanding occupations, implicitly or explicitly assuming that it is the sedentary office worker whose health has the most to gain from travelling by bicycle (Fishman, 2016). While employment in tourism and hospitality often entails a more intense kind of bodily exertion ('Out of an eight-hour shift, you are probably on your feet seven hours and 45 min' – Daniel, bartender, Brussels), it was framed by cycling commuters not in terms of an additional physical effort, but in terms of the opportunities for psychological relaxation it could offer:

> Many people have the wrong idea about our line of work. Yes, there are long shifts, 14–15 h sometimes, but it's not just being on your feet the entire time. (…) Mostly, it's not physical exhaustion so much as a mental tiredness. You have to interact with different people throughout your shift. (…) So, for the people who come by bike to work, it's not a problem to cycle home at the end of a shift, because it's a relaxing thing, a way to unwind emotionally and psychologically. (Julia, waitress, Sofia)

For the tourism and hospitality workers interviewed, the benefits of the active commute were interpreted in the context of the emotional labour they engaged in (Guerrier & Adib, 2003). To reflect this, efforts to promote cycling should go beyond the current focus on active commuting as a counterbalance to sedentary work. In addition, opportunities to improve access to shared bicycle schemes at night (when their current usage is low) could be explored by local authorities. Given the low and/or precarious incomes of many night workers, affordability measures would be required, including reduced prices after 8pm, and lowering current barriers to adoption such as prohibitive subscription and deposit fees.

Given the difficulties small employers face in providing facilities for cyclists, and the expectations placed on tourism and hospitality workers' personal presentation, shared bicycle schemes could accommodate these night commuters' needs where private bicycles do not. Thus, cycling may often be a solution for the return part of the journey only. The capacity of the shared bike to offer a solution for those who can only cycle at the end of a work shift but not before, was demonstrated by Georgina. While working as a restaurant waitress in Brussels, she took public transport on the way to work, and a Villo shared bike at night.[4] For Georgina, the location of bicycle docking stations near both origin and destination made this seemingly complex arrangement fairly convenient.

The role of employers in night-time commuting

The employers who participated in this study were not oblivious to the challenges faced by night commuters. Three-quarters of the employers surveyed reported implementing some form of measure for making employee journeys more affordable and/or convenient (on the other hand, only 57% of employees reported such measures being in place, raising questions around the extent to which all workers were made aware of mobility-related measures). One Brussels and one Sofia hotel owner specifically adjusted the night receptionists' shifts to ensure that they begin and end at times which are well served by public transport. A hotly debated topic across all interviews was the ability of employers to pay for taxis home at night. Overall, Sofia-based businesses were more likely to offer a fully paid-for taxi ride home for workers on a late shift:

> Even when we recruit people, in the advert we say that we provide free transport at the end of the evening shift. This is crucial. For the last one-and-a-half years, we've had a contract with a taxi company, and a button which is just here, in the restaurant. We press the button, and a taxi comes – you don't have to call, you don't have to say where the taxi is going. The member of staff takes a voucher from the office, gives it to the driver, and then they invoice us at the end of the month. (Vasil, restaurant owner/head chef, Sofia)

The employers who took such an approach to organising the night commute reasoned that they had a responsibility of care towards their staff, but also highlighted the strategic business importance of the subsidised taxi ride in terms of reducing staff turnover. All employers interviewed spoke of the difficulties they faced in recruiting and retaining staff, and recognised problems with night commuting as a dimension of these difficulties. Subsidising nocturnal travel was thus seen as a costly investment, but one which paid off in terms of staff loyalty.

However, a blanket scheme where all employees would be entitled to a fully paid-for taxi ride home on all occasions they worked at night, was relatively rare. In Brussels, one restaurant owner reported making a small contribution towards taxi costs on nights when shifts finished later than usual ('I give two or three euro per person, given that a taxi ride is about 10 euro'). Other employers in both Sofia and Brussels subsidised taxi costs selectively:

> We have this one waitress, she has been with us a long time. She has a small kid now, and in the evening, she wants to get back home as quickly as possible. So, we pay for her taxi. Just hers. We want to keep her. (Stanislav, restaurant owner, Sofia)

While sustainable transport research and policy have repeatedly highlighted the importance of workplace mobility planning, implementing comprehensive transport-related strategies remains out of reach for many small and medium-sized businesses, due to their scarce organisational and financial resources.[5] The approaches for dealing with the complexities of night commuting reported by the tourism and hospitality business owners were largely ad-hoc, responding to changing circumstances, often from one week to the next. The employers drew on their knowledge of their staff, and the close-knit relationships which developed in their small teams. One employer described the continuous renegotiation of night commuting arrangements not as a burden, but as an expression of their organisation's flexibility, informal communication in the workplace, and culture of cooperation:

> If you're a more junior, less experienced member of staff, you are paid less. So, we will always try and give you a lift home. This is not a formal arrangement. We simply can't afford to give people money for taxis. But we are a very small team, we all know where people live, how everyone travels here and back. If someone drives, they will offer a lift to another. We just talk to each other, we are a very close team, and we just make it work. (Anya, bar owner, Sofia)

A minority of employers felt that the expectation that they should plan night commutes with their employees, or contribute to them financially, was unfair. One Brussels-based business owner perceived employee demands for support with commuting as an erosion of a previously existent culture and work ethic. The owner of a high-end restaurant in Sofia perceived his business as a site of professional development and spoke of the many learning opportunities his employees had compared to others in the sector. For him, making one's way to work and back independently was an expression of dedication and drive which he expected from workers. From the employees' perspective, such an approach often meant that it was up to them and their colleagues to negotiate the night commute amongst themselves, as a shared challenge.

Collegial relations and collective strategies

In transport research, the commuter is often conceptualised as a solitary figure, an individual moving through space while seeking to minimise contact with others, and making travel-related decisions as a bounded independent agent (cf. Manderscheid, 2013). However, the commute can also be thought of as a socio-spatial practice which is an extension of both workplace and household relations, with individual journeys always entangled in the commitments and responsibilities linking the commuter to other people (Plyushteva & Schwanen, 2018). Similarly, the night commute, as a complex arrangement requiring continuous rethinking of constraints and strategies, was demonstrated by the accounts from Sofia and Brussels respondents to be an extension of workplace relations. On many occasions, the need to return home in a way which was affordable, convenient and safe was discussed regularly with colleagues, and ad-hoc strategies were developed on an almost-nightly basis:

> In my old job, a [female] colleague and I got into the habit of combining our socialising plans after work, so then we could walk home together – she lives very close. The more I think about it, the more I realize how much [the night commute] is an issue for us. We talk a lot about transport during shifts. (Georgiana, waitress, Brussels)

As a shared challenge, the night commute could, on one hand, initiate cooperation and mutual care. On the other hand, it acted as a way of reinforcing and sustaining existing workplace friendships and relations of care:

> One of my colleagues lives very close to me. It happens that we share a taxi on the way home; we finish at 1am, or 2am on Fridays and Saturdays. We actually got a taxi yesterday. I fancied a walk. But she was very tired. So, we got a taxi together. (Ivaylo, bartender, Sofia)

Thus, collegiality made the night commute possible or improved the experience of it, while the commute in turn became a shared concern over which colleagues got to know each other better. Several interviewees discussed the socialising with colleagues which often follows a late shift – whether in the place of work, or in bars and clubs nearby. This practice reflected the need to unwind, while also building upon the friendships and relations which staff in closely located venues often engaged in. For one Sofia-based respondent, boundaries between acting as an employee and as customer could become easily blurred – after finishing his work as a bartender, he would visit a friend's bar for a drink, and occasionally lend a hand if the venue was busy. His experience was one which problematised any facile distinction between the consumers of night-life and those who produce it. The blurring of categories of work and leisure in tourism and hospitality is well-documented, and accounts such as this one are important in interpreting employees' position in urban nightlife as multifaceted and open-ended (Bianchi, 2000; Guerrier & Adib, 2003). Applying these insights to the nocturnal commute points to the need to complicate rigid categories of journey purpose, and recognise the extent to which work and leisure-related mobilities in the urban night cannot be disentangled.

When it comes to transport policy and planning, night transport services and discourses around them should expand their focus from the consumption of night leisure to include the range of journeys which take place at night.

The practices of support and cooperation which emerged during and after late shifts highlighted the ways in which night commuters were never simply victims of circumstance. On one hand, research participants acknowledged the extensive demands placed upon them to deliver particular kinds of experiences to night revellers, while dealing with low and unpredictable incomes, limited employer support, and ebbs and flows in custom which required constant adjustment of their personal daily and nightly rhythms. On the other hand, they did not simply endure nocturnal work and travel, but organised them as active agents, involved in intricate collective strategies and negotiations. Nevertheless, as demonstrated above, these negotiations often represent a significant psychological, physical and financial burden. They do not reduce the need for local decision-makers and individual employers to better account for the mobility needs of those whose labour sustains night-time cities as destinations for leisure and sites of economic activity.

Discussion and conclusion

This paper undertook an examination of the barriers which hospitality and tourism workers encounter when they have to commute at night, and the resources they draw upon in addressing them. It sought to highlight the specific challenges which result from the nature of tourism and hospitality work, the ways in which they amplify the limitations of night-time transport in cities, and the practices which employers and workers developed in tackling these challenges. Rather than a stable and predictable arrangement, the nocturnal commute of workers in Sofia and Brussels emerged as an undertaking made possible through continuous renegotiation, involving adjusting choices of transport mode and extending workplace cooperation and care into practices of organising mobility. Both employers and employees emphasised the significant financial burden of commuting at night, with persistent ambiguities and tensions around who should shoulder this cost. Being able to share nocturnal journeys with colleagues was important in making them safer and more affordable, but shift patterns and the cost of a taxi at night meant alternatives had to be found, often at short notice.

This discussion contributes to the growing body of research on the mobilities of tourism and hospitality, which to date has focussed predominantly on the mobilities of tourists, and the role of international migrant workers in these sectors. By focussing on the intricate negotiations involved in everyday (and everynight) mobilities, I have sought to demonstrate the ways in which seemingly mundane journeys are part of sustaining tourism and hospitality services. In light of growing research and policy calls to re-conceptualize the so-called 'night-time economy' in more inclusive terms (Hadfield, 2015; Roberts & Eldridge, 2007; Schwanen, van Aalst, Brands, & Timan, 2012), I have argued that mobility at night presents both workers and employers with difficult questions around affordability, safety, fatigue and workplace relations. The nocturnal commute of tourism and hospitality workers thus foregrounds issues which can often remain invisible to dominant conceptions of the urban night as a timespace of freedom, leisure and escape from daytime pressures. For city authorities and businesses aiming to develop nocturnal cities as leisure destinations, the night commute points to the particular mobility needs of those working the late shift: on one hand, they cannot be assumed to coincide with the mobilities of night revellers; on the other hand, they problematise rigid distinctions between travelling for work and for leisure.

In exploring the night-time commute of tourism and hospitality workers, this paper has aimed to address the scarcity of research on night-time mobilities more broadly. Transport and mobilities research has overwhelmingly focussed on daytime urban travel, with peak hour congestion prioritised in both data collection and transport planning. Public transport provision at night is absent or scarce in many European cities, with services often limited to Friday and Saturday nights. As a result, night-time commuters can face uncertainty, high costs, as well as concerns about personal safety. While this paper examined commuting in the tourism and hospitality sectors, the nocturnal mobilities of cleaning, maintenance, health, transport and security services, to name a few, remain largely unknown. Further research is needed to make the particular barriers to nocturnal mobility visible, and to ensure that they are reflected in urban transport policy agendas.

Notes

1. While some daytime public transport services continued to run until approx. midnight, field observations revealed very infrequent services after 11pm in most study locations.
2. Collecto operates between 11pm and 6am every night of the week. It requires that the user books their trip at least half an hour in advance, and boards the taxi at one of the designated Collecto stops across the capital region. A central dispatch point ensures taxi rides are shared by up to 3 passengers with similar routes, keeping costs lower. The Collecto service is paid for by the local authority, and provided by sub-contracted taxi companies.
3. Noctis is the brand under which the Brussels transport authority operates its night buses.
4. Villo is a shared bicycle scheme operated by the Brussels Capital Region in partnership with a private company. It was launched in 2009, and as of 2014 there were 346 stations where bicycles could be rented or dropped off. A number of pricing options are available, but a 150-euro deposit is required before using the scheme. As of April 2018, a similar scheme is in development in Sofia.
5. In Belgium, only companies with 100 or more employees are required to have a formal workplace mobility plan in place, which excludes the majority of tourism and hospitality businesses. No such legal requirement exists in Bulgaria.

Disclosure Statement

No potential conflict of interest was reported by the author.

Funding

This work was supported by the European Union's Horizon 2020 research and innovation programme under the Marie Skłodowska-Curie grant agreement No 703047.

ORCID

Anna Plyushteva http://orcid.org/0000-0002-0701-1525

References

Anable, J. (2005). 'Complacent car addicts' or 'aspiring environmentalists'? Identifying travel behaviour segments using attitude theory. *Transport Policy*, *12*(1), 65–78. doi:10.1016/j.tranpol.2004.11.004

Banister, D. (2008). The sustainable mobility paradigm. *Transport Policy*, *15*, 73–80.

Baum, T. (2007). Human resources in tourism: Still waiting for change. *Tourism Management*, *28*(6), 1383–1399. doi:10.1016/j.tourman.2007.04.005

Baum, T., Kralj, A., Robinson, R. N. S., & Solnet, D. J. (2016). Tourism workforce research: A review, taxonomy and agenda. *Annals of Tourism Research*, *60*, 1–22. doi:10.1016/j.annals.2016.04.003

Bianchi, R. V. (2000). Migrant tourist-workers: Exploring the 'contact zones' of post-industrial tourism. *Current Issues in Tourism*, *3*(2), 107–137. doi:10.1080/13683500008667869

Bissell, D. (2010). Passenger mobilities: Affective atmospheres and the sociality of public transport. *Environment and Planning D: Society and Space*, *28*, 270–289.

Bissell, D. (2011). Thinking habits for uncertain subjects: Movement, stillness, susceptibility. *Environment and Planning A*, *43*, 2649–2665.

Bissell, D. (2018). *Transit life: How commuting is transforming our cities.* Cambridge, MA: The MIT Press.

Brandeleer, C., Ermans, T., Hubert, M., Janssens, I., Lannoy, P., Loir, C., & Vanderstraeten, P. (2016). Cahiers de l'Observatoire de la mobilité de la Région de Bruxelles-Capitale: Cahier 5: Le partage de l'espace public en Région de Bruxelles-Capitale. Brussels: Bruxelles Mobilité.

Collectiu Punt 6. (2017). *Nocturnas: The everyday life of women nightshift workers in the Barcelona metropolitan area.* Barcelona: Collectiu Punt 6.

Corvellec, H., & O'Dell, T. (2012). Demanding hosts and ungrateful guests – the everyday drama of public transportation in three acts and academic prose. *Culture and Organization*, *18*(3), 231–249. doi:10.1080/14759551.2011.634195

Duncan, T., Scott, D. G., & Baum, T. (2013). The mobilities of hospitality work: An exploration of issues and debates. *Annals of Tourism Research*, *41*, 1–19. doi:10.1016/j.annals.2012.10.004

Edensor, T. (2007). Mundane mobilities, performances and spaces of tourism. *Social & Cultural Geography*, *8*(2), 199–215. doi:10.1080/14649360701360089

Eldridge, A., & Roberts, M. (2008). A comfortable night out? Alcohol, drunkenness and inclusive town centres. *Area*, *40*(3), 365–374.

England, K. (1996). Mothers, wives, workers: The everyday lives of working mothers. In K. England (Ed.), *Who will mind the baby? Geographies of child care and working mothers* (pp. 109–122). London: Routledge.

Eurostat. (2018). *Regional tourism statistics.* Retrieved from http://ec.europa.eu/eurostat/data/database

Fishman, E. (2016). Cycling as transport. *Transport Reviews*, *36*(1), 1–8. doi:10.1080/01441647.2015.1114271

Gerike, R., de Nazelle, A., Nieuwenhuijsen, M., Panis, L. I., Anaya, E., Avila-Palencia, I., … on behalf of the PASTA consortium. (2016). Physical activity through sustainable transport approaches (PASTA): A study protocol for a multicentre project. *BMJ Open*, *6*(1), e009924. doi:10.1136/bmjopen-2015-009924

Gonzales, E. J., & Daganzo, C. F. (2013). The evening commute with cars and transit: Duality results and user equilibrium for the combined morning and evening peaks. *Transportation Research Part B: Methodological*, *57*, 286–299. doi:10.1016/j.trb.2013.06.009

Guerrier, Y., & Adib, A. (2003). Work at leisure and leisure at work: A study of the emotional labour of tour reps. *Human Relations*, *56*(11), 1399–1417. doi:10.1177/00187267035611006

Hadfield, P. (2015). The night-time city. Four modes of exclusion: Reflections on the urban studies special collection. *Urban Studies*, *52*(3), 606–616. doi:10.1177/0042098014552934

Hall, C. M. (2005). Reconsidering the geography of tourism and contemporary mobility. *Geographical Research*, *43*(2), 125–139. doi:10.1111/j.1745-5871.2005.00308.x

Hollands, R., & Chatterton, P. (2003). Producing nightlife in the new urban entertainment economy: Corporatization, branding and market segmentation. *International Journal of Urban and Regional Research, 27*(2), 361–385.

Ioannides, D., & Zampoukos, K. (2018). Tourism's labour geographies: Bringing tourism into work and work into tourism. *Tourism Geographies, 20*(1), 1–10. doi:10.1080/14616688.2017.1409261

Jordhus-Lier, D., & Underthun, A. (2015). *A hospitable world?: Organising work and workers in hotels and tourist resorts*. London; New York: Routledge.

Kwan, M.-P. (1999). Gender, the home-work link, and space-time patterns of nonemployment activities. *Economic Geography, 75*(4), 370–394.

Liempt, V., & Aalst, V. (2015). Whose responsibility? The role of bouncers in policing the public spaces of nightlife districts. *International Journal of Urban and Regional Research, 39*(6), 1251–1262. doi:10.1111/1468-2427.12320

Lucas, K. (2012). Transport and social exclusion: Where are we now? *Transport Policy, 20*, 105–113. doi:10.1016/j.tranpol.2012.01.013

Manderscheid, K. (2013). Criticising the solitary mobile subject: Researching relational mobilities and reflecting on mobile methods. *Mobilities*, 1–32. doi:10.1080/17450101.2013.830406

McDowell, L., Batnitzky, A., & Dyer, S. (2007). Division, segmentation, and interpellation: The embodied labors of migrant workers in a greater London hotel. *Economic Geography, 83*(1), 1–25.

Middleton, J. (2009). 'Stepping in time': Walking, time, and space in the city. *Environment and Planning A, 41*, 1943–1961.

Monaghan, L. F. (2002). Regulating 'unruly' bodies: Work tasks, conflict and violence in Britain's night-time economy. *British Journal of Sociology, 53*(3), 403–429. doi:10.1080/0007131022000000572

National Statistics Institute. (2018). Population by regions, areas, registered address and gender. Retrieved from http://www.nsi.bg/bg/content/2975/%D0%BD%D0%B0%D1%81%D0%B5%D0%BB%D0%B5%D0%BD%D0%B8%D0%B5-%D0%BF%D0%BE-%D0%BE%D0%B1%D0%BB%D0%B0%D1%81%D1%82%D0%B8-%D0%BE%D0%B1%D1%89%D0%B8%D0%BD%D0%B8-%D0%BC%D0%B5%D1%81%D1%82%D0%BE%D0%B6%D0%B8%D0%B2%D0%B5%D0%B5%D0%BD%D0%B5-%D0%B8-%D0%BF%D0%BE%D0%BB

O'Dell, T. (2009). My soul for a seat: Commuting and the routines of mobility. In E. Shove, F. Trentmann, & R. Wilk (Eds.), *Time, consumption and everyday life: Practice, materiality and culture* (pp. 85–98). Oxford: Berg.

Plyushteva, A. (2016). Commuting with new transport infrastructure: Change, cost and comfort on the Sofia metro extension (PhD Thesis). University College London, London, UK.

Plyushteva, A., & Schwanen, T. (2018). Care-related journeys over the life-course: Thinking mobility biographies with gender, care and the household. *Geoforum, 97*, 131–141.

Roberts, M., & Eldridge, A. (2007). Quieter, safer, cheaper: Planning for a more inclusive evening and night-time economy. *Planning Practice and Research, 22*(2), 253–266. doi:10.1080/02697450701584410

Schlich, R., & Axhausen, K. W. (2003). Habitual travel behaviour: Evidence from a six-week travel diary. *Transportation, 30*(1), 13–36.

Schwanen, T. (2007). Gender differences in chauffeuring children among dual-earner families. *The Professional Geographer, 59*(4), 447–462. doi:10.1111/j.1467-9272.2007.00634.x

Schwanen, T., Banister, D., & Anable, J. (2011). Scientific research about climate change mitigation in transport: A critical review. *Transportation Research Part A: Policy and Practice, 45*(10), 993–1006. doi:10.1016/j.tra.2011.09.005

Schwanen, T., Lucas, K., Akyelken, N., Cisternas Solsona, D., Carrasco, J.-A., & Neutens, T. (2015). Rethinking the links between social exclusion and transport disadvantage through the lens of social capital. *Transportation Research Part A: Policy and Practice, 74*, 123–135. doi:10.1016/j.tra.2015.02.012

Schwanen, T., van Aalst, I., Brands, J., & Timan, T. (2012). Rhythms of the night: Spatiotemporal inequalities in the nighttime economy. *Environment and Planning A, 44*(9), 2064–2085. doi:10.1068/a44494

Shaw, R. (2018). *The nocturnal city*. Abingdon, Oxon; New York, NY: Routledge.

StatBel. (2018). Structure of the Population. Retrieved from https://statbel.fgov.be/en/themes/population/structure-population#panel-12

Van Malderen, L., Jourquin, B., Thomas, I., Vanoutrive, T., Verhetsel, A., & Witlox, F. (2012). On the mobility policies of companies: What are the good practices? The Belgian case. *Transport Policy, 21*, 10–19. doi:10.1016/j.tranpol.2011.12.005

Vladimirov, Z., & Vedar, O. (2016). Risk factors, preventive measures, compensation and job satisfaction of employees from night establishments in Bulgaria. *Contemporary Management Practices, 2016*(9), 102–110.

Waitt, G., Harada, T., & Duffy, M. (2017). 'Let's have some music': Sound, gender and car mobility. *Mobilities, 12*(3), 324–342. doi:10.1080/17450101.2015.1076628

Zografsky, A. (2018, July 9). How can we improve night transport. Retrieved from http://spasisofia.org/bg/%D0%BA%D0%B0%D0%BA-%D0%B4%D0%B0-%D0%BF%D0%BE%D0%B4%D0%BE%D0%B1%D1%80%D0%B8%D0%BC-%D0%BD%D0%BE%D1%89%D0%BD%D0%B8%D1%8F-%D1%82%D1%80%D0%B0%D0%BD%D1%81%D0%BF%D0%BE%D1%80%D1%82.html

Strangers in the night: nightlife studies and new urban tourism

Adam Eldridge

ABSTRACT

This paper draws together recent scholarship from the study of urban tourism and nightlife. Though studies of urban tourism do not always specifically address nightlife, and likewise studies of the night and nightlife do not always examine tourism, both bodies of research overlap in important ways. Concerns about commercialisation, gentrification, displacement, and urban change are to be found in both bodies of research. However, while the study of urban tourism typically recognises the erasure of the host / guest binary and seeks to destabilise the notion of who is a tourist or stranger, studies of nightlife often rest on a much clearer distinction between who belongs and who does not. An argument proposed here is that while the host / guest, tourist / non-tourist binary is perhaps reconfiguring, the night and nightlife spaces reinstate these binaries in various ways. This paper thinks through debates about tourists and residents in the night, focusing in particular on questions of belonging, place identification and gentrification through night-time uses.

Introduction

A point of departure for this paper concerns the closure of Fabric, a nightclub located in central London opposite the historic Smithfield meat market. Opened in 1999, the club was closed by licensing authorities in September 2016 due, in part, to the drug-related deaths of two patrons. A social media campaign soon followed and the closure was commented upon by a number of spokespeople, including London's Mayor, Sadiq Khan (Baines, 2016). Though Fabric eventually re-opened, in light of wider concerns about the closure of a number of nightlife venues, especially local pubs, live music and LGBTQ+ venues, news of the club's closure fed into the wider perception that London's nightlife was under threat (Campkin & Marshall, 2017; Furedi, 2015).

It was while attending a conference on nightlife in Berlin two months later that the fallout of Fabric's closure became clear to me. As well as being a frequent topic of discussion, a fundraising event had been organised for conference delegates to help cover Fabric's legal costs. As someone who lives very close to Fabric, it was both heartening but also unexpected to find people, almost 1000 km away, organising fundraising events and expressing concern about the closure of my 'local' club. Something I did not admit to at the time, however, was that despite my proximity and interest in nightlife I had actually never been to Fabric. Talking to my neighbours about the closure revealed similarly ambiguous feelings; one felt it was a shame for young people but was more concerned about the recent closure of one of our local parks popular with dog walkers. My neighbours downstairs, who regularly attended the club, were disappointed but took a pragmatic view that clubs opened and closed regularly.

Talking to my neighbours and people in Berlin about Fabric raised many of the issues explored in this paper. In particular, I am interested in the relationship 'hosts and guests' have to space, the role of tourists and residents in making cities, and, more broadly, place attachment and belonging (Cohen & Cohen, 2012). The discussion draws on recent scholarship from the field of new urban tourism (Hannam, 2009; Novy & Huning, 2009; Novy, 2018); understood here as the blurring of binaries that traditionally characterise tourism and how it is understood: local versus resident, touristic versus local places, producers versus consumers, and quotidian versus tourist experiences. The paper examines these debates but does so through the lens of the night, drawing on a related but slightly different set of terms and concerns. I am specifically interested in the night here as a contributing factor that shapes how tourists and locals alike might experience the urban landscape after dark and how the night contours our feelings of belonging or otherwise. That tourists might feel a different sense of attachment or belonging to a place than locals, and that we might experience these feelings differently in the night, points to a series of questions about how 'mobility and moorings' (Brenner, 2004; Cresswell, 2010) are as much temporal as spatial concepts. How, or to what extent, the night and nightlife more generally disrupts or reinstates the binary of tourist and non-tourist, belonging or strangeness, is of particular interest. After a brief discussion of debates about the night in the UK, and the dominance of the conflict model, the paper moves on to explore how tourism at night blurs or reinforces various distinctions through which nightlife is understood, lived and imagined. Guiding the discussion is the question of to what extent the night shapes different relations, entanglements and intensities of being and 'territorial belonging' (Gustafson, 2009) in the nocturnal city.

Conflict and nightlife

A now well-established argument about nightlife is that it can lead to conflict between residents and nightlife consumers, with the latter often presumed to be 'not local'. Since at least the 1990s, the expansion of nightlife in British town and city centres, particularly alcohol-related establishments, has been marked by concerns about anti-social behaviour, binge drinking, the commercialisation of leisure, and over-development (Chatterton & Hollands, 2002; Hadfield, 2006; Lovatt, 2017; Roberts & Turner, 2005). Better management of what has become known as 'the night-time economy' has been a goal of various tiers of government from the late 1990s onwards (Roberts, 2004), with different policies seeking to both champion and restrain further development. While the liberalisation of licensing through England and Wales' Licensing Act (2003) and the recent appointment of night tsars / mayors in London and Manchester can be understood more in terms of the former, other initiatives such as closing venues and other licensed public spaces, regulating hours, and allowing local residents to comment on and object to licensing decisions have sought to regulate, inhibit or at the very least better orientate its growth.

A second narrative characterising the literature on nightlife in British cities is its ambiguous relationship to regeneration and gentrification (Hobbs, Winlow, Hadfield, & Lister, 2005; Roberts & Eldridge, 2009). In the 1980s and 1990s, nightlife was seen in some quarters as a tool for reversing de-industrial decline in urban centres (Lovatt & O'Connor, 1995; Stevenson, 2018). This could be achieved by encouraging workers to remain in town centres at the end of their work day, promoting bars, restaurants, and other 'quarters', encouraging housing development in urban centres – often alongside emerging nightlife precincts – and staging events after dark (Heath, 1997; Hughes, 1999). An active and vibrant nightlife was understood to act as a draw for further investment, recalling both Florida's (2005) work in articulating creativity and vibrancy with economic growth, and, Lloyd's (2010) work on urban neo-bohemia in promoting a 'grit as glamour' urban aesthetic.

The link between nightlife, regeneration and gentrification was always slightly ambiguous, however. On the one hand, nightlife, including aspirational bars, restaurants and clubs in new mixed-use developments, could be seen to encourage further growth by helping to market and 'brand' the nocturnal city. Night-time leisure venues and vibrant and diverse streetscapes would attract not just new homeowners, renters, investors, and students, but would also open up other supply chains, such as take-away venues, taxi companies, and various businesses supplying bars and clubs (Roberts & Eldridge, 2009). On the other hand, noise, binge drinking and the presence of some forms of nightlife deemed to be anti-social could negatively impact on property prices and detract from further development. As explored elsewhere (Eldridge, 2009, see also Thurnell-Read, 2012), hen and stag parties were one such example where groups of presumably loud and drunken people were actively discouraged lest they alarm families or other nightlife consumers and investors. The promotion of British cities after dark has not been neutral, in other words, but instead has encouraged creative, aspirational, safe and well-managed venues and streets (Haydock, 2014). This has not only led to accusations of nightlife being used to gentrify certain neighbourhoods and displace existing residents but also highlights how some users, local or otherwise, have been privileged over others.

Central to much of the literature on British nightlife has thus been the theme of conflict, be that between residents and consumers, nightlife providers, local authorities and other stakeholders, new and established residents, promoters, and those resisting the over-commercialisation of youth spaces and cultures. Much of these debates about nightlife have been framed in terms of alcohol and binge drinking, especially in regards to students and young working-class people, but wider debates have also been raised more generally about who then has access to urban centres at night and what that might mean for 'night citizenship' (Gwiazdzinski, 2014; Plyushteva, 2009). What we want our nocturnal cities to be, how they might be more accessible and inclusive, and who are they for are chief, albeit not always clearly articulated, concerns.

Some of these arguments echo in studies of conflicts surrounding tourism and unsurprisingly there has been a significant crossover between the two bodies of literature (Nofre et al., 2017; Nofre, Giordano, Eldridge, Martins, & Sequera, 2018). Commercialisation, rising land values, residential and commercial displacement (Cocola-Gant, 2018), and attempts to maintain local services run by and for local residents are concerns that have also informed research into sites where tourists and locals are more likely to cross over. These connect with debates about the nocturnal city in important ways. As extensively documented in cities such as Barcelona (Nofre et al., 2018), Amsterdam (Pinkster & Boterman, 2017) and Lisbon, (Nofre et al., 2017), nightlife spaces play a central role in many of the conflicts that tourism is believed to cause; noise, gentrification, and challenges to local identity. Again, it is a question of who cities are for and who has the power to shape their present and future that drives many recent studies examining the entanglement of tourism and nightlife in European cities, especially areas which have only recently began to experience the expansion of tourism. These particular neighbourhoods, often in transition areas 'off the beaten track' (Maitland & Newman, 2009), echo the concerns explored thus far. Füller and Michel, in their study of Berlin's Kreuzberg, for example, note that 'the demand of the new urban tourism for "authentic' experiences results in an interest in precisely the amenities, the retail and entertainment infrastructure that city residents also prefer … [however] the transformation of working-class pubs and mom-and-pop grocery stores into high-priced coffee shops are only the most visible signs of such conflicts' (2014, p. 1314). Like other formerly working-class neighbourhoods such as Shoreditch, London (Pappalepore, 2010), Brooklyn, New York (Hae, 2011), or parts of Istanbul, Turkey (Eder & Öz, 2015), Kreuzberg has been subject to a wave of galleries, bars, and restaurants opening in quick succession, with students and urban tourists drawn to the area as Berlin has become the new 'epicentre of cool' (Füller & Michel, 2014). The effect of this on local residents, including commercial or residential displacement through rising rents and neighbourhood change, has dominated local discourse and media reports about nightlife in Kreuzberg (Füller & Michel, 2014).

Numerous other studies have painted a similar picture (Nofre & Eldridge, 2018), one where tourism at night is entangled with broader debates about gentrification, the intensification of commercialism at night, displacement of local residents and the leisure or retail spaces that serve them, as well as more quotidian concerns such as the ability to sleep, feelings of safety and belonging, or simply having somewhere familiar in which to socialise after dark. There is important and considerable reach to this line of inquiry, especially in terms of how it draws attention to the ways that urban tourism shapes cities, cities shape urban tourism, and the extent to which nightlife potentially intensifies both. Indeed, a contention here is that conflicts resulting from increased tourism might become especially resonant at night. Local or neighbourhood bars and restaurants changing their offer, revellers disturbing the sleep patterns of nearby residents, apartments being taken over by investors or residents renting them through peer-to-peer networks, guests hosting parties in such apartments and disturbing other residents, and the transformation of local shops and services into new tourism-led bars and clubs are all especially relevant.

More recently, however, tourist scholars have critically examined the terms upon which such debates often rest (Novy, 2018). For some time now, it has been argued that the host /guest binary is not necessarily stable or fixed, and 'the tourist' is not a stable identity category or easy to define (Cohen & Cohen, 2012). While binaries such as the resident /nightlife consumer or resident /tourist still informs much of the critical work explaining conflict in the night, the following section examines how the night might further complicate the ways we understand hosts, guests, tourists and place attachment.

New urban tourism

As discussed above, conflict remains a dominant way of understanding nightlife, especially between residents and nightlife consumers. In no way does this discussion seek to minimise the debates raised by this body of work. Leisure is one site through which our identities and connections to place are constituted and experienced (Rojek, 2005), and something as simple as the closure of our local pub, or finding it transformed for the pleasure of new or even non-residents can lead to a sense of dislocation from our own leisure spaces. That said, the discussion now turns to consider how an emphasis on conflict between residents, tourists and nightlife consumers more generally might overly rely on an 'us and them' logic, a logic which has been challenged by urban tourism scholars. Much of this work, under the rubric of new urban tourism, has sought to challenge the binaries through which tourism is typically framed. If, in the past, tourists were understood as primarily consumers of spaces designated as touristic and experienced the urban in ways that clearly differentiated them from locals, more recent work has sought to question whether this model continues to work. Instead, and as the discussion now explains, new technologies, desires, experiences and mobilities mean tourists not only experience cities 'like a local' and 'off the beaten track', but play an equally important role in co-constituting places (Dirksmeier & Helbrecht, 2015; Giovanardi, Lucarelli, & Decosta, 2014; Novy, 2018).

Füller and Michel (2014) recite what is now a fairly well accepted argument in the study of urban tourism; a desire for more 'authentic' or more 'ordinary' experiences leads the new urban tourist 'off the beaten track' (Maitland & Newman, 2009) into typically transitioning areas. These areas are often working class, associated with ethnic minorities, and might be the setting for cultural, creative and experimental venues and atmospheres (Novy & Huning, 2009). In contrast to the traditional tourist bubbles, where tourists have desires, demands and interests distinct from residents (Lim & Bouchon, 2017), a desire for ordinary or more everyday 'creative' experiences sees today's urban tourists venture into such areas in the hope of having a less 'touristy' experience. For Pappalepore et al. these gentrified or gentrifying 'microgeographies … challenge traditional views of tourism consumption by blurring the boundaries between tourists, day visitors and residents' (2014, p. 228). An important consequence of this is an entanglement of resident and tourist practices and interests. As Lim and Bouchon (2017) characterise it, tourists engage in behaviours coded as touristy, such as visiting galleries or shopping, while also engaging in practices coded as residential or local. This might include shopping in local grocery stores, eating in local cafes and restaurants, attending local farmers markets, using laundrettes, or participating in local events.

The breaking down of the host / guest binary is further enabled through the use of digital technologies and social media and how these allow us to enact a closer or at least *different* relationship to place (Hannam, Butler, & Paris, 2014). For Bock, such technologies challenge the 'information divide' (2015, p. 14; cited in Novy, 2018, p. 428), whereby being 'a local' no longer determines who has expertise about a local area. How this might be quantified or qualified is difficult to determine and depends greatly on how belonging and place attachment are understood (Lewicka, 2011), as is discussed later, but the anecdote used at the beginning of this paper, my own ambiguous relationship to Fabric nightclub, is one such example where people not from my immediate neighbourhood were much more concerned than I was about the club's closure. My experience in Berlin some months later confirmed that I also knew far less about the case and what the club offered than nightlife consumers in a different country. This anecdote very much echoes Lewicka's (2011) extensive review of articles about people-place relations, where it was found that academic interest in people-place attachments was increasingly concerned with places that were not necessarily our permanent homes. That is, if earlier research into place attachment emphasised stability, temporal connections or 'rootedness', more recent work on tourist-place relations has uncovered similarly meaningful relations to place by people otherwise deemed as mobile or 'not local'. While not discounting the importance of ancestral or cultural connections to place, tourists can develop 'deeply meaningful' (Kianicka, Buchecker, Hunziker, & Müller-Böker, 2006) relations to place – an argument that has been supported to various degrees by studies of second homeowners (Roca, 2016), sports tourists (Brown, Smith, & Assaker, 2016), and in both regional and urban environments (McIntyre, Williams, & McHugh, 2006).

Garcia's (2016) study of techno-tourism in Berlin is one such example of this body of work and starts with an opposing anecdote to my own about British clubbers who visit Berlin's techno-clubs every third or fourth week of the month. This is a unique example, but it illustrates the extent to which nightlife tourists will work to belong, and, by extension, break down the resident-tourist binary. Garcia's interviewees plan their visits in advance, stay at friend's apartments, make use of existing networks, and draw on their considerable cultural knowledge of DJs, clubs and affordable places to eat. Their resulting sense of attachment to the clubs of Berlin, enabled through digital technologies and the ways knowledge circulates distinct from geography, is best encapsulated by one of Garcia's interviewees claiming: 'This is my local club, even if I have to get on a plane to come here' (Garcia, 2016, p. 285).

With this in mind, defining who is the local or tourist becomes challenging. In a period of increased mobility, for some at least (Cresswell, 2010), and having access to considerable knowledge about a place before even venturing there, means our relationship or sense of attachment is not easily determined by or reducible to residency. The ERASMUS students referred to in Nofre et al.'s (2018) study, the repeat visitor (Freytag, 2010), the business person who regularly stays in the same hotel, or the techno-tourist (Garcia, 2016), share a different, yet no less important sense of attachment and investment in a place where they do not permanently live. In large global cities such as Berlin, London or New York, marked as they are by tourism, residential churn, and diversity, it becomes further challenging to define exactly who belongs or who is a 'local'. As Dirksmeier and Helbrecht suggest, 'urban tourists are just strangers among strangers' (2015, p. 278).

Strangers at night?

Thus far it has been argued that literature on the night overlaps with literature on tourism, especially in transitioning areas popular with those seeking a more authentic experience off the beaten track. Whether it is a concern with gentrification, commercialisation, locals versus newcomers, or anxieties about urban change and gentrification of night-time and other spaces, these are all entangled with wider processes linked to urban tourism and, more generally, post-industrial urban transformations (Novy, 2018).

Moreover, in the study of urban tourism there is an established argument that the host /guest or tourist /non-tourist binaries are increasingly challenged by various mobilities and information technologies.

This attempt to rethink and challenge such binaries might function differently in regards to the study of the night, however. Indeed, in studies of nightlife debates about noise, morality, and who an area's leisure provision is actually for tend to rest on much more familiar understandings of the resident, the local, and the stranger. The question that motivates the second part of this paper is whether, in the modern global, cosmopolitan city, night studies similarly needs to rethink the host /guest, or resident /non-resident binary or whether the ways the night is lived, imagined and plays a role in constituting our identities renders those relations of proximity and sense of attachment more complex. While not discounting the argument that in contemporary cities we are perhaps 'strangers amongst strangers', is this useful in thinking through what happens at night and the ways we experience cities after dark?

Iveson explains that the city has long been a privileged site of analysis for cosmopolitanism, where traditional kinship or tribal ties give way to new and 'uncertain ... identifications' (2006, p. 70). In modern global cities, all urban dwellers are understood in terms of 'strangeness', effectively flattening any claims to space, territory or subjecthood. If we are all strangers, all mobile and all constantly arriving and departing (Iveson, 2006), then any claims to the host–guest binary further breaks down, taking with it any greater claim to the urban realm. That is, if modern urban life is indeed about strangers living amongst strangers all cobbled together, then the tourist becomes simply one further stranger negotiating their way through an intensely complex world of familiarity, difference, and greater or lesser intensities of belonging. Attempts to enforce a local /non-local binary, be it through claims to public space, 'my' bar rather than 'theirs', gated suburbs, or restrictive policies determined by 'locals' is antithetical to this particular discourse of modern global cities.

This account represents a rather extreme understanding of the de-territorialized, mobile, urban citizen, however. As well as the need to be mindful of history, power, and privilege (Uteng & Cresswell, 2016), and the ways that mobilities might privilege masculinist models of home and travel (Morris, 1988), a problem with the idea that we are all strangers thrown together is that it flattens out actual hierarchies of difference and how those differences are lived, managed and made meaningful in everyday or structural ways. As Iveson suggests, 'Any demand that all urban inhabitants adopt a cosmopolitan openness to others will have a fundamentally different meaning for weak groups than it will for those who have voluntarily fortified themselves in enclaves of privilege' (2006, p. 81). We also need to be careful, as Hannam (2009) and others have stated, of romanticising or privileging the figure of the nomad, mindful of how this can represent a new form of power, prestige and cosmopolitan cultural capital (Rofe, 2003). Flying to Berlin every month to go clubbing, or knowing all the latest restaurants and bars to visit in any other modern, cosmopolitan city depends on forms of knowledge and financial power that are not universally shared.

The belief that we are all strangers, or that we are all swept up into a mobile and fluid world, also obscures different intensities of place attachment. The study of place attachment is broad and vexed (Altman & Low, 1992; Lewicka, 2011) but has been defined as 'the emotional link formed by an individual to a physical site that has been given meaning through interaction' (Milligan, 1998, p. 2; cited in Harmon, 2006). While it is beyond the scope of this discussion to enter into the myriad ways the concept is understood and deployed, the issue of how the night might contour our sense of attachment has been less examined in the literature. Temporality is a key theme in the study of place attachments, be that in the form of childhood homes, nostalgia for past landscapes, or the citation of annual rituals and events (Altman & Low, 1992). How attachment to place changes by time of the day remains somewhat obscure, however, but central to understanding how the night further contours our moorings and mobilities. Place attachments, and the sites that we wish to 'moor' (Debarbieux, 2014) ourselves to, change from day to night, effecting different relations of familiarity, connection or disconnection, in turn producing different understandings of who or who is not a stranger.

To draw this out, it is important to explain how the night is understood here. The night is often figured as a transgressive space free of everyday rules and norms (Lovatt & O'Connor, 1995; Palmer, 2000; Williams, 2008). In many respects, the ways the night is commonly understood mirrors common sense understandings of tourism – a time when everyday constraints are suspended and we are free of domestic and work schedules or commitments (Edensor, 2007). Nightlife spaces can similarly 'undo' the day and challenge common hierarchies of power while disrupting all that is articulated with the day; the quotidian, work, labour, oppressive norms, and the commercial realm of business (Williams, 2008). The night, more generally, is often further articulated with notions of creativity, cultural change, experimentation and resistance. Discourses of morality and crime continue to structure our understanding of the night, but it is equally understood as a space for friendship, sociability, and leisure. It is perhaps for these reasons that the commercialisation of leisure spaces or punitive restraints on night-time leisure is so strongly resisted. Leisure 'is the birthplace of the self, of the realization of one's own nature pursued purely for its own sake' (Cohen-Gewerc & Stebbins, 2013, p. 106). When a local pub personally associated with friendship and localism is closed or subject to change it challenges more than one's identity but all that the night is supposed to be; freedom, leisure, and escape from everyday structures, including the imperatives of commercialism.

This is, once again, a highly romanticised view of the night and just as tourism can reproduce existing forms of privilege, the choices available to us at night are highly gendered, raced, classed, and aged. To even have access to the night depends on forms of privilege, such that developing place attachments might not even be possible. Nonetheless, for those who actively engage in nightlife, the night is a key site for the production and reproduction of (sub)cultural capital (Thornton, 1996) and therefore the experience of belonging and attachment. Nightlife spaces such as clubs, restaurants and bars rest on familiarity with rules, norms and expectations which we are expected to know (May, 2014; Thornton, 1996). Door policies are an obvious example of this but so too are the subtler web of powers and potential pleasures and embarrassments that shape our engagement with the venues, streets and public spaces we access after dark. Even in non-mainstream or alternative venues which disrupt and resist quotidian hierarchies, knowledge and know-how in the form of embodied and symbolic capital plays an essential role in enjoying, let alone being able to access, these spaces. Failure to grasp how nightlife spaces constitute identities and operate to include and exclude means failing to recognise why they become so important, so fiercely protected, and often generate precisely the same binaries new urban tourism scholars have challenged: us and them, local or stranger, fitting in or being an outsider. A local club or pub can thus be lived and imagined as integral to individual and community histories, desires, representations and indeed place attachments.

While nightlife spaces might, therefore, be reliant on familiarity and forms of symbolic and cultural capital which, by their nature, exclude others from participating, this does not necessarily mean geography is the key determinant of who has access. Returning to Garcia (2016), know-how is not predicated on being a local resident. In this case, any presumed link between geography and community is challenged by the evidence of people travelling vast distances to their 'local club', and the ways that information and knowledge about where to go, what to do and how to consume a city after dark circulate beyond locally specific networks. Garcia's interviewees demonstrate that it is not our physical proximity to a venue that counts, but rather a much more complicated sense of ownership, place attachment, and the desire to belong in ways that are not determined by our usual place of residence and might not even be relevant in daytime hours. In effect, the us and them binary often seen in accounts of tourism at night are not simply about where we call home, but how we fit into the prevailing culture on offer.

Further to this, while accounts of tourism and mobilities might privilege that sense of the mobile stranger, the night is not simply about strangers meeting strangers. It is also about seeing and making friends, mixing with people we already know (whether they live here or there), and developing forms of kinship and identity. All sites are heavily invested with specific cultural meanings and can be central to the constitution of our identities. The night, however, bound up as it is with pleasure, cultural capital, identity, friendship, for some their youth, for others alienation and fear, and a broad range of other emotions (Hubbard, 2016) is strongly associated with a very complex set of desires and practices of belonging. This is not to privilege nightlife spaces as any more affective or crucial to our identity, but to suggest that, perhaps, the ways we experience and are anchored within and consume cities is different in the night compared to the day.

In these terms, we might think of place attachment and identification at night beyond the broad descriptor of 'the neighbourhood' or 'city', and who has claim to them, but more in a micro sense of clubs, bars, and other leisure sites (Hernández, Hidalgo, Salazar-Laplace, & Hess, 2007) where belonging is less spatially determined and is enabled instead by complex 'conduits' of mobility, especially taste, capital (Cresswell, 2010, p. 24) and tribes (Maffesoli, 1996). Indeed, these might well be spaces where 'locals' feel no place attachment to such sites at all, despite living in the neighbourhoods in which they are located. As Hernández et al. (2007) explain:

> one person could be attached to a place but not be identified with it (i.e. someone who likes to live in a place and wants to remain there but does not feel that this place is part of their identity; at least not their main place identity) and vice versa; someone could have a high personal identity with a place and not a high place attachment (for example, to feel that one belongs to a place but prefers not to live there) (Hernández et al., 2007, p. 311).

The ways that place attachments, by locals or otherwise, intersect with mobilities and tourism is therefore complex, and further contoured by the ways that the night is such an important timespace through which identities and identifications are understood, imagined, and lived. Understanding the night and nightlife spaces as crucial to constituting identity means we are not simply strangers rubbing up against other strangers, but are often trying precisely to not be strangers and find and perform instead a sense of commonality and shared cultural forms of belonging, meaning, identity and 'moorings'. If this results in or constitutes an us and them logic, or a sense of belonging and not belonging, then that is precisely because nightlife is one of many important spaces for the affirmation of identity and inclusion across local or multiple sites.

Concluding thoughts

The picture I have analysed here is a complex one. Because the night is so bound up with pleasure and identity for some people, change can be strongly resisted. When a local bar, restaurant or club becomes gentrified, whether that be attributed to tourism or the more general processes of urban middle-classification, it can be strongly resisted in us and them terms. This and the ways that much of the conflict between nightlife is framed as tourists or strangers vs residents depends on and reproduces a logic which urban tourism scholars have rightfully sought to question. Us and them at night are not necessarily about geography, or how long we have lived in a place, however, but whether we fit into the dominant culture on offer, whether we know the music, wear the right clothes, understand the menu, or are confident in negotiating the familiar and the strange. Taste in music, food, or fashion and the confidence and cultural capital to enter and enjoy nightlife venues play a greater role in our sense of attachment than where we reside in regards to how we emotionally attach ourselves to certain night-time spaces – be those attachments fixed or fleeting.

The (dis)attachments we might have to our cities after dark, the clubs, bars or streets, are already highly contoured by gender, class, age, and our own symbolic and cultural capital. Participation and a sense of belonging or not belonging are therefore heightened at night. In rapidly gentrifying areas in particular, new shiny bars or restaurants might deliberately exclude some long-term residents in favour of a new, more mobile, more privileged clientele, instigating a further wave of tension between newer and older residents, the mobile or the constrained. A caveat to this, however, is the ways that local policies and laws do continue to allow local residents to shape their local night-time provision. Local residents with the skills or knowledge necessary to engage with local licensing decisions can participate in shaping licensing hours, the provision of outdoor seating, or even when a bar closes, and those decisions can impact on people living thousands of miles away who feel no less intimate or invested in the nightlife provision of that area. This power might not translate into actually using those spaces or indeed feeling welcome, however.

The study of urban tourism, especially in regards to transitioning areas off the beaten track, has been instrumental in opening up the binaries through which urban life is typically conceived and has sought to question such terms as the stranger, local, tourist, and resident. With it, questions about place attachment, identification and who makes places have been similarly raised. But a contention here is that the night complicates feelings of belonging or disconnection. We might get angry about noise that would not otherwise bother us during the day. We might be more sensitive to bars and clubs where there are unfamiliar codes of behaviour or indeed where we find our own 'tribe' has moved on or been excluded. Leisure spaces are intensely social and therefore different to the laundrettes, public transport, street markets or local shops we might share with non-residents or tourists in the day. This is not to suggest all places are not similarly affective or woven into and constitutive of our identities, but the night, because it is already entangled with wider concerns about exclusion or inclusion, identity and being with our tribe, complicates the ways we do and consume the urban realm.

A point that becomes clear here, following Gustafson (2001, 2009, 2014) is that mobility does not necessarily erase being attached to specific places. The night-time reveller might well be a stranger on one scale but thoroughly embedded at the level of a club or bar. Likewise, the local resident, deeply rooted and dependent on local connections, might feel alienated and distanced from their nightscape, be that through traditional axes of exclusion such as gender or age, or through the social changes prompted by gentrification, urban change and indeed mobilities. Nightlife scholars, while increasingly sensitive to the ways the night excludes and includes, would no less benefit, however, from drawing on scholars of urban tourism in considering the complex entanglements of place and identity that the night produces, while cognisant of breaking down the notion of us and them.

Disclosure statement

No potential conflict of interest was reported by the author.

ORCID

Adam Eldridge http://orcid.org/0000-0002-9262-8983

References

Altman, I., & Low, S. M. (Eds.). (1992). *Place Attachment.* London: Plenum Press.

Baines, J. (2016, September 2). London Mayor Sadiq Khan opens up about fabric and London nightlife in new interview. *Thump.* https://thump.vice.com/en_uk/article/pg8zxg/london-mayor-sadiq-khan-opens-up-about-fabric-and-london-nightlife-in-new-interview

Bock, K. (2015). The changing nature of city tourism and its possible implications for the future of cities. *European Journal of Futures Research, 3*(1), 1–8.

Brenner, N. (2004). *New state spaces: Urban governance and the rescaling of statehood.* Oxford: Oxford University Press.

Brown, G., Smith, A., & Assaker, G. (2016). Revisiting the host city: An empirical examination of sport involvement, place attachment, event satisfaction and spectator intentions at the London Olympics. *Tourism Management, 55*, 160–172.

Campkin, B., & Marshall, L. (2017). *LGBTQ+ cultural infrastructure in London: Night venues, 2006–present.* https://www.ucl.ac.uk/urbanlab/docs/LGBTQ_cultural_infrastructure_in_London_nightlife_venues_2006_to_the_present.pdf

Chatterton, P., & Hollands, R. (2002). Theorising urban playscapes: Producing, regulating and consuming youthful nightlife city spaces. *Urban Studies, 39*(1), 95–116.

Cocola-Gant, A. (2018). Tourism gentrification. In L. Lees & M. Phillips (Eds.), *Handbook of gentrification studies* (pp. 281–293). Cheltenham: Edward Elgar Publishing.

Cohen, E., & Cohen, S. A. (2012). Current sociological theories and issues in tourism. *Annals of Tourism Research, 39*(4), 2177–2202.

Cohen-Gewerc, E., & Stebbins, R. A. (2013). *Serious leisure and individuality.* Montreal: McGill-Queen's Press.

Cresswell, T. (2010). Towards a politics of mobility. *Environment and Planning D: Society and Space, 28*(1), 17–31.

Debarbieux, B. (2014). Rootedness – anchoring – mooring: Reviving metaphors. *L'Espace Geographique, 43* (1), 68–80.
Dirksmeier, P., & Helbrecht, I. (2015). Resident perceptions of new urban tourism: A neglected geography of prejudice. *Geography Compass, 9*(5), 276–285.
Edensor, T. (2007). Mundane mobilities, performances and spaces of tourism. *Social & Cultural Geography, 8*(2), 199–215.
Eder, M., & Öz, Ö. (2015). Neoliberalization of Istanbul's nightlife: Beer or Champagne? *International Journal of Urban and Regional Research, 39*(2), 284–304.
Eldridge, A. (2009). "Drunk, fat and vulgar": the problem with hen parties. In C. Kevin (Ed.), *Feminism and the body: Interdisciplinary perspectives* (pp. 194–211). Newcastle: Cambridge Scholars Publishing.
Florida, R. (2005). *Cities and the creative class.* Oxon: Routledge.
Freytag, T. (2010). Visitor activities and inner-city tourist mobility: The case of Heidelberg. In: J. A. Mazanec & K. W. Wöber (Eds.), *Analysing international city tourism* (pp. 213–226). Vienna: Springer.
Furedi, F. (2015). *Forward into the night.* Night Time Industries Association. http://www.ntia.co.uk/wp-content/uploads/2015/06/Forward-into-the-Night-Report.pdf?567959
Füller, H., & Michel, B. (2014). 'Stop being a tourist!' New dynamics of urban tourism in Berlin-Kreuzberg. *International Journal of Urban and Regional Research, 38*(4), 1304–1318.
Garcia, L. M. (2016). Techno-tourism and post-industrial neo-romanticism in Berlin's electronic dance music scenes. *Tourist Studies, 16*(3), 276–295.
Giovanardi, M., Lucarelli, A., & Decosta, P. L. E. (2014). Co-performing tourism places: The "Pink Night" festival. *Annals of Tourism Research, 44*, 102–115.
Gustafson, P. (2001). Roots and routes: Exploring the relationship between place attachment and mobility. *Environment and Behavior, 33*(5), 667–686.
Gustafson, P. (2009). Mobility and territorial belonging. *Environment and Behaviour, 41*(4), 490–508.
Gustafson, P. (2014). Place attachment in an age of mobility. In L. C. Manzo & P. Devine-Wright (Eds.), *Place attachment: Advances in theory, methods and applications* (pp. 37–48). Oxon: Routledge.
Gwiazdzinski, L. (2014). Pleading for the right to the city's night. Night manifesto. Seeking Citizenship 24 h, Invisiveis Producoes, 204–219.
Hadfield, P. (2006). *Bar wars: Contesting the night in contemporary British cities.* Oxford: Oxford University Press.
Hae, L. (2011). Dilemmas of the nightlife fix: Post-industrialisation and the gentrification of night-life in New York city. *Urban Studies, 48*(16), 3449–3465.
Hannam, K. (2009). The end of tourism? Nomadology and the mobilities paradigm. In J. Tribe (Ed.), *Philosophical issues in tourism* (pp. 101–113). Clevedon: Channel View.
Hannam, K., Butler, G., & Paris, C. M. (2014). Developments and key issues in tourism mobilities. *Annals of Tourism Research, 44*, 171–185.
Harmon, D. (2006). *People, places, and parks: Proceedings of the 2005 George Wright Society Conference on parks, protected areas, and cultural sites.* Hancock: TheGeorge Wright Society. [http://www.georgewright.org/0525harmon.pdf]
Haydock, W. (2014). The 'civilising' effect of a 'balanced' night-time economy for 'better people': Class and the cosmopolitan limit in the consumption and regulation of alcohol in Bournemouth. *Journal of Policy Research in Tourism, Leisure and Events, 6*(2), 172–185.
Heath, T. (1997). The twenty-four hour city concept – a review of initiatives in British cities. *Journal of Urban Design, 2*(2), 193–204.
Hernández, B., Hidalgo, M. C., Salazar-Laplace, M. E., & Hess, S. (2007). Place attachment and place identity in natives and non-natives. *Journal of Environmental Psychology, 27*(4), 310–319.
Hobbs, D., Winlow, S., Hadfield, P., & Lister, S. (2005). Violent hypocrisy: Governance and the night-time economy. *European Journal of Criminology, 2*(2), 161–183.
Hubbard, P. (2016). The geographies of 'going out': Emotion and embodiment in the evening economy. In L. Bondi & J. Davidson (Eds.), *Emotional Geographies* (pp. 131–148). Abingdon: Routledge.

Hughes, G. (1999). Urban revitalization: The use of festive time strategies. *Leisure Studies, 18*(2), 119–135.

Iveson, K. (2006). Strangers in the cosmopolis. In J. Binnie, J. Holloway, C. Young, & S. Millington (Eds.), *Cosmopolitan urbanism* (pp. 70–86). Oxon: Routledge.

Kianicka, S., Buchecker, M., Hunziker, M., & Müller-Böker, U. (2006). Locals' and tourists' sense of place. *Mountain Research and Development, 26*(1), 55–64.

Lewicka, M. (2011). Place attachment: How far have we come in the last 40 years? *Journal of Environmental Psychology, 31*(3), 207–230.

Lim, S. E. Y., & Bouchon, F. (2017). Blending in for a life less ordinary? Off the beaten track tourism experiences in the global city. *Geoforum; Journal of Physical, Human, and Regional Geosciences, 86*, 13–15.

Lloyd, R. (2010). *Neo-bohemia: Art and commerce in the postindustrial city.* New York: Routledge.

Lovatt, A. (2017). The ecstasy of urban regeneration: Regulation of the night-time economy in the transition to a post-Fordist city. In J. O'Connor & D. Wynne (Eds.), *From the margins to the centre* (pp. 141–168). Oxon: Routledge.

Lovatt, A., & O'Connor, J. (1995). Cities and the night-time economy. *Planning Practice & Research, 10*(2), 127–134.

Maffesoli, M. (1996). *The time of the tribes: The decline of individualism.* (D. Smith,Trans.). London: Sage.

Maitland, R., & Newman, P. (2009). Developing world tourism cities. In R. Maitland & P. Newman (Eds.), *World tourism cities. Developing tourism off the beaten track* (pp. 1–21). London: Routledge.

May, R. A. B. (2014). *Urban nightlife: Entertaining race, class, and culture in public space.* London: Rutgers University Press.

McIntyre, N., Williams, D., & McHugh, K. (2006). *Multiple dwelling and tourism: Negotiating place, home and identity.* Oxfordshire: Cabi.

Milligan, M. J. (1998). Interactional past and potential: The social construction of place attachment. *Symbolic Interaction, 21*(1), 1–33.

Morris, M. (1988). At Henry Parkes motel. *Cultural Studies, 2*(1), 1–47.

Nofre, J., & Eldridge, A. (2018). *Exploring nightlife: Space, society and governance.* London: Rowman & Littlefield.

Nofre, J., Giordano, E., Eldridge, A., Martins, J. C., & Sequera, J. (2018). Tourism, nightlife and planning: Challenges and opportunities for community liveability in La Barceloneta. *Tourism Geographies, 20*(3), 377–396.

Nofre, J., Sánchez-Fuarros, I., Martins, J. C., Pereira, P., Soares, I., Malet-Calvo, D., … López Díaz, A. (2017). Exploring nightlife and urban change in Bairro Alto, Lisbon. *City & Community, 16* (3), 330–344.

Novy, J. (2018). 'Destination' Berlin revisited. From (new) tourism towards a pentagon of mobility and place consumption. *Tourism Geographies, 20*(3), 418–442.

Novy, J., & Huning, S. (2009). New tourism areas in the "New" Berlin. In R. Maitland, & P. Newman (Eds.), *World tourism cities* (pp. 87–108). London: Routledge.

Palmer, B. D. (2000). *Cultures of darkness: Night travels in the histories of transgression [from medieval to modern].* New York, NY: Monthly Review Press.

Pappalepore, I. (2010). *Tourism and the development of 'creative' urban areas: evidence from four non-central areas in London* (Doctoral dissertation, University of Westminster).

Pappalepore, I., Maitland, R., & Smith, A. (2014). Prosuming creative urban areas. Evidence from East London. *Annals of Tourism Research, 44*, 227–240.

Pinkster, F. M., & Boterman, W. R. (2017). When the spell is broken: Gentrification, urban tourism and privileged discontent in the Amsterdam canal district. *Cultural Geographies, 24*(3), 457–472.

Plyushteva, A. (2009). The right to the city and struggles over urban citizenship: Exploring the links. *Amsterdam Social Science, 1*(3), 81–97.

Roberts, M. (2004). *Good practice in managing the evening and late night economy: A literature review from an environmental perspective.* London: Office of the Deputy Prime Minister.

Roberts, M., & Eldridge, A. (2009). *Planning the night-time city.* Abingdon: Routledge.

Roberts, M., & Turner, C. (2005). Conflicts of Liveability in the 24-hour city: Learning from 48 hours in the life of London's Soho. *Journal of Urban Design*, *10*(2), 171–193.
Roca, Z. (2016). *Second home tourism in Europe: Lifestyle issues and policy responses.* London: Routledge.
Rofe, M. W. (2003). 'I want to be global': Theorising the gentrifying class as an emergent élite global community. *Urban Studies*, *40*(12), 2511–2526.
Rojek, C. (2005). *Leisure theory: Principles and practice.* Basingstoke: Palgrave Macmillan.
Stevenson, D. (2018). Feminism and its places: Women, leisure and the night-time economy. In *The Palgrave handbook of feminism and sport, leisure and physical education* (pp. 557–569). London: Palgrave Macmillan.
Thornton, S. (1996). *Club cultures: Music, media, and subcultural capital.* Middletown, CT: Wesleyan University Press.
Thurnell-Read, T. (2012). Tourism place and space: British stag tourism in Poland. *Annals of Tourism Research*, *39*(2), 801–819.
Uteng, T. P., & Cresswell, T. (2016). *Gendered mobilities.* Oxon: Routledge.
Williams, R. (2008). Night spaces: Darkness, deterritorialization, and social control. *Space and Culture*, *11*(4), 514–532.

Nocturnal ritual activities in tourist development of pilgrimage cities

Marie-Hélène Chevrier

ABSTRACT

Dusk and dawn are important times in Catholic sacred places. Some shrines are famous for their nocturnal activities, including torchlight processions, prayer vigils, and late-night or dawn services. For pilgrims, a day does not necessarily end at nightfall and often begins before dawn. In pilgrimage cities, both religious and tourist authorities have renewed traditional nocturnal rituals, which are often promoted as intensely spiritual and unique tourist experiences. The discussion below asks how these nocturnal practices differ from the current offer in terms of night tourism. Nocturnal life, as currently understood, applies mainly to capital cities and is essentially based upon festive practices and leisure, providing what might then be seen as 'artificial' pleasures. This paper, based on fieldwork conducted in fifteen Catholic shrines (in Europe and Mexico), focuses on three examples: the nocturnal Marian procession of Lourdes (France), the case of Lyon Festival of Lights, and a recent project called 'The 25th Hour' conducted by the Association of the Shrine Towns of France. These examples are used to examine the link between tourist experiences, debates about authenticity, and the night, and considers nocturnal ritual practices as powerful assets for promoting a new type of cultural night tourism in both smaller and world cities.

Introduction

Among the most visited places of the world are many Catholic shrines; Our Lady of Guadalupe in Mexico (20 million visitors per year), Aparecida in Brazil (12 million), Lourdes in France (6 million), Fatima in Portugal, and Rome (4.5 million). Special events of a religious nature, such as processions, also draw many visitors to smaller, more modest towns which are often far away from famous tourist cities. These sacred places and ritual practices are part of religious tourism[1] which represents, according to the *World Tourism Organization (UNWTO)*, 600 million national and international voyages a year. This is further demonstrated by the flourishing of 'spiritual tours' in travel agencies and by the creation of an international congress on 'Tourism and Pilgrimage' by the *UNWTO* in 2014. Not all 600 million visitors are pilgrims or believers, however. According to a survey conducted between 2013 and 2016, 72% of visitors to sanctuaries come for religious reasons, 8% out of curiosity, 17% out of cultural interest and 3% for other motives (Chevrier, 2016).

A key point developed here is that visiting sacred places is typically linked to a quest for authenticity, with that 'authenticity' believed to reside in the visitors' practices as much as in the actual places. Nocturnal ritual practices, especially in Christiandom, give great importance to the alternation of day and night which assumes a strong symbolic significance. Within the Christian tradition, 'darkness and the night had long served as powerful metaphors' (Koslofsky, 2011, p. 10). The first act of Creation reported in the Book of Genesis is dividing the light from darkness. In Christian faith and theology, as in many others, the light is associated with God and particularly to Christ, who referred to himself as 'the light of the world' (John 8, 12). When He died on the Cross, 'darkness came over the whole land' (Luke 23, 44) and the lighting of the Pascal candle, during the Easter vigil on Holy Saturday at dusk, is symbolic of His Resurrection and a climax of the liturgical year. Other nocturnal ritual practices (torchlight processions, vigils and night prayers) are similarly marked by symbolic meaning: the victory of life over death and light over darkness.

These ritual nocturnal practices contrast with the current development strategies of many nighttime economies which instead are geared towards relaxation and entertainment. This paper, in contrast, takes a different view of the night and asks how we might overcome the predominant economic approach through which it is increasingly understood. Firstly, the literature review demonstrates how little attention has been paid to the link between authenticity and the night thus far, even though authenticity is a key element of the tourist experience. The methodology section explains the qualitative data collected in the three case studies and their associated sacred practices. The final section explores how the extension of tourism space–time is enabled by the promotion of these specific nocturnal practices.

Experience, authenticity, emotions, sacred and the night

An economic understanding of nightlife has become the predominant focus, not only in research, but also for public authorities and town planners. In June 2014, for example, the French Ministry of Foreign Matters announced the creation of five Centres of Tourist Excellence focused on visitor's expectations. One of these Centres was dedicated to 'Night Tourism'. A few months later, the Ministry published a report including 22 proposals to develop nocturnal life for tourism. This development strategy was based on the twenty-four-hour city concept (Heath & Strickland, 1997) and promoted the development of a French night-time economy (NTE). It was inspired by programmes inaugurated in the early 1970s in Rome, and then developed in Berlin, Barcelona and London; these three cities being cited as examples to follow (Ministry of Foreign Affairs and International Development, 2014). The report pointed out that French nocturnal life already offers 'a multitude of activities' such as concerts, theatres, clubs, bars, shops, and restaurants, etc. and thus situated the night as the 'realm of play, of socialisation, of encounter and of evasion' (Lovatt & O'Connor, 1995, p. 130). The Ministry has sought to further transform the night into a festive extension of the day, abolishing the natural dichotomy between day and night-time uses and practices.

Apart from an inevitable array of problems associated with the development of a night-time economy such as pollution (noise and light), security and anti-social behaviour, (Gwiazdzinski, 2002, 2008; Roberts & Eldridge, 2009), these proposals for night-life display three distinctive issues for night management. Firstly, this festive conception of night-life is highly selective and is reserved for certain types of people. Studying the practices of the French population, Espinasse highlights that 'undeniably, *the night belongs to the youth*'. A majority of people going out at night are indeed aged 20–29, whereas during the 'extreme ages' of life – childhood and old age – the night is 'forbidden' (Espinasse, Gwiazdzinski, & Heurgon, 2005, p. 81). Similarly, the activities promoted imply a financial cost (spectacles, restaurants, nightclubs) that are affordable primarily for the upper classes, especially as theatres, bars and restaurants are generally concentrated in central and gentrified neighbourhoods.

Secondly, this type of programme overlooks nightlife in small towns; only regional and capital cities possess the means to generate such permanent festive activity. Indeed, Evans (2011) underlines that night-time events are part of the competition between cities at a national and supranational level.

Finally, two of the most important peculiarities of the night are forgotten, even denied, by the night-time economy as it has been promoted by recent French policy: silence and obscurity. The night-time economy is based on the existence of artificial light converting the night into an extension of the day, forgetting that some nocturnal practices such as religious ones (torchlight processions, illuminations) and light festivals, are based on an equilibrium between silence and obscurity on one hand and specific sounds and lights on the other. In the document of the French Ministry for instance, there is no mention of Lyon Festival of Lights which is one of the world's largest light festivals. Darkness being necessary for their success, these events raise the need to reconsider the relation between light and dark: neither dichotomy nor continuity, but complementarity (Edensor, 2013).

Currently, urban night-time policies privilege recreational and diversional modes of tourist experiences where tourism is viewed as a search for pleasure (Cohen, 1979a). This festive framing of the night and its implications have come to constitute the main focus for researchers on nightlife across Europe. The night is mostly considered as a time for entertainment and leisure: for tourists after a day of attending what could be called 'tourist matters' (guided tours, long walks or beach activities, for instance) as well as for residents after a day of work. It does not consider specific lived experiences or the fact that a significant part of the tourist experience is a quest for 'meaning' and 'authenticity'.

When night tourism is involved in touristic practices it is essential to refer to the notion of tourist experience, defined by Cohen as 'the inner state of the individual, brought about by something which is personally encountered, undergone or lived through' (Cohen *in* Jafari, 2000). This 'something' is a set of stimuli both 'objective' (weather, temperature, etc.) and 'subjective' (tourist brochures, historical or cultural meaning of a place, etc.). Through these experiences, the visitor might have a feeling of finding an 'authentic self'. such as through becoming one of the individuals she came to meet. Wang and Gnoth remind us that 'Tourist experiences are not just about seeing sights but are more importantly about performing places' (*in* Jafari & Xiao, 2016). The desire to have a tourist experience, and thus do more than just sightseeing, can be understood as a desire or a quest for authenticity, understood here as both an emotion and a product of the visitors' judgement or subjectivity (Cohen, 1979b; Wang, 1999). According to MacCannell (1976), in tourist places, a 'true' or 'objective' authenticity must be distinguished from a 'staged' authenticity which does not fool or mislead tourists. This distinction might be convincing in the case of a type of tourism essentially based on the tourist gaze, which compares the visitor with a spectator, but it does not necessarily help when looking at tourism from the 'tourist as performer' angle. One vivid example is when the visitor becomes a participant such that the quality of the experience might not only depend on the 'objective' authenticity. Instead, thinking of the tourist as a performer, staged authenticity evolves towards 'existential' authenticity (Wang, 1999) or 'experiential' authenticity (Daniel, 1996) which implies seeking to understand the Other and temporarily adopting his or her logic. Existential authenticity is thus both intrapersonal (related to the ability of human beings to be transformed by an experience) and interpersonal (related to collective experiences and the formation of 'touristic communitas') (Moufahim & Lichrou, 2018; Wang, 1999).

The desire to have a tourist experience and the related quest for authenticity can be understood through the example of tourists visiting sacred places and taking part in various religious events which leads to a form of existential alterity. Shrines and religious practices provide emotional experiences, a 'hot authenticity', different from the objectivity of 'cool authenticity' (Selwyn, 1996). Indeed, whatever the actual authenticity of a sanctuary[2], the visitor's perceptions are determined by what she is inclined to believe and what she is ready to accept as true, with these depending on personal beliefs. In sacred places, authenticity not only depends on the qualities of the place but also the characteristics of the visitors: their beliefs, their activities in the place, their identity and the practices in which they participate. The visitors' characteristics combine with those of the place creating 'theoplacity' (Belhassen, Caton, & Stewart, 2008), but while the theoplacity framework details the general concept of authenticity relevant to the specific case of religious places, it does not consider factors such as the weather or the time of the day. Given how important the quest for authenticity is in contemporary tourism, and supposing that the environment in which the tourist experience takes place is an additional stimulus (Wang & Gnoth *in* Jafari & Xiao, 2016), the little interest taken by researchers in the relations between night, authenticity and the sacred is surprising. Indeed, existing literature, particularly the Romantic and then Surrealist Movements, contributed to create a *topos*, associating night with mystery, the sublime, and special feelings such as liberty and fear (Bonte, 2017). But reflecting on the affective and sensual dimension of the night emerged only recently in geography. According to Edensor's works (2012-2015), for example, light transforms space and choices made in terms of lighting (artistic uses of light) are able to create specific atmospheres which contribute to 'defamiliarize' familiar places and 'solicit feelings and activates sensations' (Edensor, 2015, p. 130). What is of particular interest here is that it is precisely the *staging* of the night which seems to generate emotions. The activities are based on the very essence of the night, allowing the visitors an other, more sensorial, experience of space.

Among nocturnal activities, ritual practices such as processions and events derived from some light festivals are particularly interesting cases, especially in relation to MacCannell's condemnation of staged authenticity. A rite is: 'a ceremonial, a cult, a movement prescribed by the liturgy of a religion' (Albert & Piette *in* Azria & Hervieu-Léger, 2010) and is characterized by a specific sequence of movements and often requires some unusual garments and tools. Nocturnal rituals are like a staging of the night and all it is supposed to mean. While performed as a show of faith, they are coupled with sincerity, which could be perceived as genuine by the visitor (believer or not) because of the commitment of the performer. Thus, the spiritual, the ritual and the authentic are strongly bound, which provides an importance to sacred places and religious events in tourism market. The fact that the ritual torchlight processions of Lourdes and Lyon Festival of Lights attract visitors in their thousands leads us to wonder about the visitors' motivation. What are they looking for, that they do not find in the recreational dimension of the night tourism developed in major cities? The cases explored below, of nocturnal practices linked to religious rituals, demonstrate that authenticity is a powerful element for the development of experiential, experimental and existential modes of tourist experiences at night.

Methodological framework: a comparative study, qualitative data

Context: three different kinds of French night events

Torchlight procession of Lourdes

Among the numerous Catholic shrines, Lourdes is the most famous. This small town (14,300 inhabitants) located in the French department of High-Pyrenees surprisingly holds a national position right behind Paris in terms of hotel establishments, with a capacity of 30,355 beds. The sanctuary of Notre-Dame de Lourdes, created after the apparitions of the Virgin Mary to Bernadette Soubirous in 1858, attracts approximately 6 million visitors per year and is a very important economic resource for this region. Lourdes is particularly renowned for its torchlight processions. Each day, at dusk (9.00pm), from Easter to the end of October, pilgrims and visitors gather at the Grotto, pray the Rosary meditating on the Holy Mysteries, sing the *Ave Maria* and perform a circumambulation of the esplanade of the shrine following a pre-established winding itinerary. The esplanade has been designed to accommodate 40,000 visitors and is frequently fully occupied, especially during the high season of the pilgrimage from April to the end of October.

The second event discussed below, Lyon Festival of Lights, marks the tourist climax of the year in Lyon, which is the third most important city of France. Since 1999, around the 8th of December, iconic buildings such as the cathedral, the City Hall, as well as some or less known places such as tramway tunnels or parks, are enlightened and animated for four nights (45 installations in 2017[3]), attracting between 2 and 4 million visitors. According to Lyon tourist office, approximately 65 international delegations are represented during those four days, the event being a 'global magnet' for light festival lovers and planners (Edensor, 2014).

The 25th Hour of French Shrine Towns is the third example discussed. The Association of Shrine Towns in France, created in 1996, gathers seventeen French cities where a partnership is formed between their shrine and tourist office (Figure 1). Great shrine towns (for example Lourdes or Mont-Saint-Michel) share their experience with smaller ones (such as Thierenbach) and some itineraries are developed between different towns of the network. The Association launched a night event in 2014, 'The 25th Hour', which takes place every two years on the last Saturday of October to coincide with the changing from EST. Each town proposes an 'unusual' visit by night: torchlight processions recalling a pilgrimage (Lourdes, La Salette, Cotignac); historical reconstructions of the medieval pilgrim arrival with torches (Chartres); and a walk with lanterns (Guebwiller; Saint-

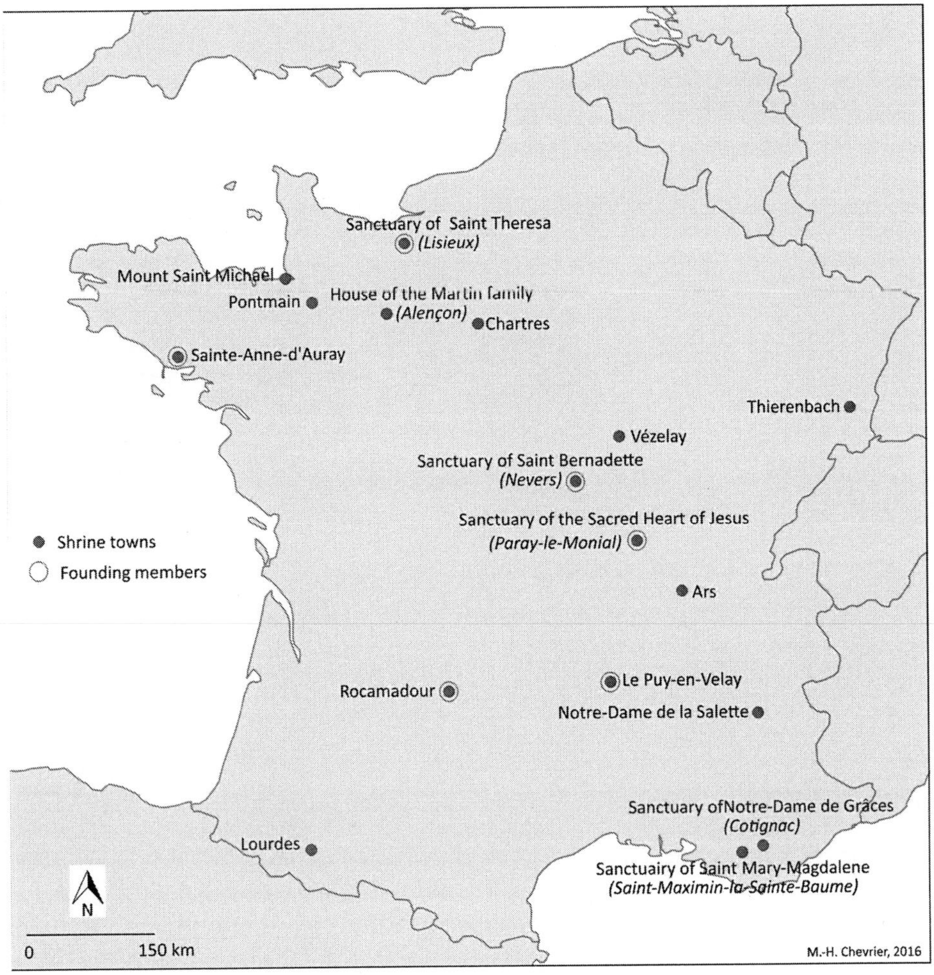

Figure 1. The members of the Association of Shrine Towns in France in 2016 (M.-H. Chevrier, 2016).

Maximin-la-Sainte-Baume, Vézelay). In these towns, itineraries are chosen to be symbolic: the ways of Santiago (Chartres, Rocamadour); an itinerary of the Saints or the apparitions (Nevers, Lisieux, Pontmain); and guided tours of basilicas with unusual animations calling upon the sense of hearing (voices muttering stories in the crypt of Sainte-Anne-d'Auray).

Data collection and methodology

The cases of Lourdes, Lyon and some of the shrine towns of the Association presented here were part of a comparative study of fifteen Catholic shrines in Western Europe and Mexico City (Figure 2). They were selected according to their canonical status, date of foundation, influence, origin of the rite, the importance of tourism and pilgrimage and balance between these two types of practices. The research worked from the assumption that secularization favoured the opening of these religious sanctuaries to tourism (Chevrier, 2016). Therefore, the visitors were increasingly not only pilgrims but also tourists with or without religious motives. The tourists' spatial practices differ from the pilgrims' and

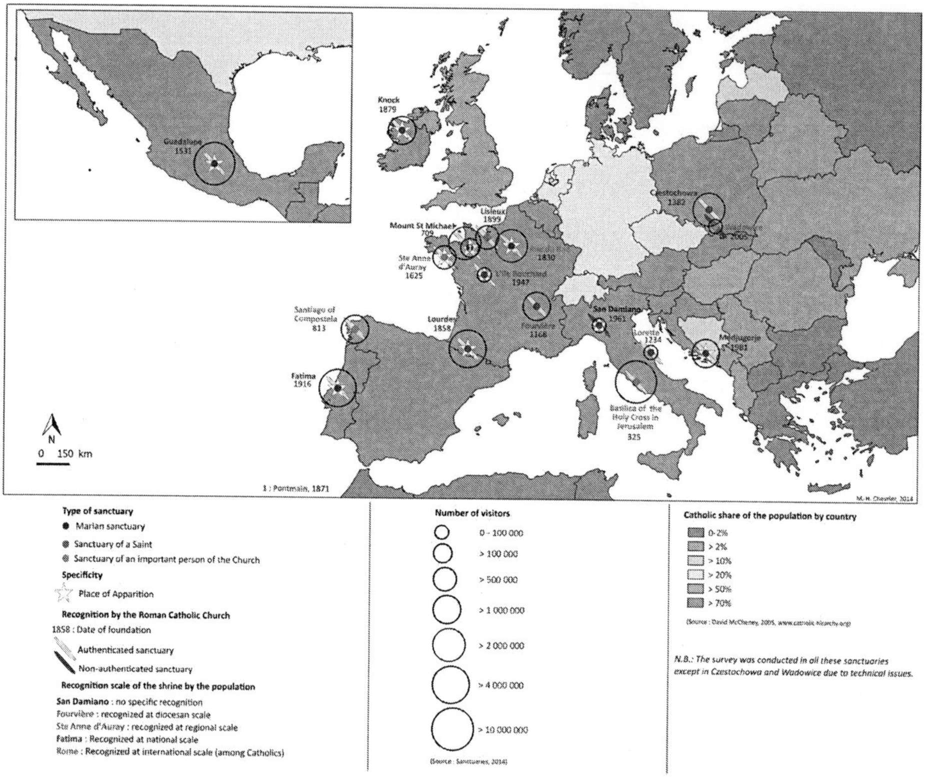

Figure 2. Map of the sanctuaries where the survey was conducted (M.-H. Chevrier, 2016).

the resulting clash has led to many repercussions for the management of the shrines. First, a phase of non-participant observation was conducted in each shrine, in preparation for a survey. The observations focused on the spatial practices and behaviour of visitors during their strolls in the sanctuaries. The observer made a list of their itineraries, stops, and gestures (religious or secular). These observations as well as the survey were conducted during the opening hours of the shrines and at various times of the year (including feast days). A survey then aimed to better understand the visitors' perceptions and representations of sacred spaces. It was conducted in thirteen of the European shrines (the survey was not authorized by the Association managing San Damiano), using direct questionnaires. Six hundred and fifty-six visitors (about fifty per shrine) with 43 different nationalities answered 19 questions. Most of the questions were closed ones, but others were semi-open, inviting the visitors, for example, to classify different locations of the sanctuary by order of importance according to their feelings, or to give the three best words to describe the shrine. The survey was completed by informal conversations with the respondents, often eager to share their feelings. Finally, 27 semi-structured interviews with those involved in the management, development and promotion of the sanctuaries were conducted: priests, ecclesiastical rectors of the sanctuaries, tourist offices and travel agencies employees, and volunteers in charge of pilgrimage.

The aim of the study was to examine the epistemological question of the dichotomy between pilgrims and tourists, and to measure the impact of the secularization of society on the sacred space via these two spatial practices (pilgrimage and tourism). Nocturnal ritual practices were not the central subject of the survey, however, the importance of night-time experiences in sacred places clearly emerged from the informal conversations with visitors and the interviews with people in charge.

The paper now turns to examine how nocturnal ritual practices are experienced as 'authentic' by the visitors and exploited as such by the authorities; with a missionary perspective for the ecclesiastical authorities and with a commercial perspective for the tourist and public authorities. The paper also discusses how the night is rehabilitated as a space–time for tourism thanks to the sensual and sacred dimension of these practices.

Findings: making darkness a special place for the tourist experience

The creation of a new landscape at night

At night, beautiful landscapes and sights – powerful attractions for tourists (Urry, 1990; MacCannell, 1976) – disappear. The gaze is constricted. 'Night versions' of emblematic landscapes (major cities and famous sites such as Niagara Falls) were created thanks to electric light. But these artificial lights were installed to conjure the day away, not to use and bring out the characteristics of the night.

Conversely, all three events examined here do not try to erase the night but magnify it and give it meaning. Illustrating Edensor's work, another relationship between light and dark is presented here: neither the ancestral dichotomy, nor the modern continuity of day blurring into night, but a much greater sense of complementarity through a specific aesthetic of place (Edensor, 2013). Nocturnal ritual activities create new landscapes and new sights, based on the balance between light and dark. The sight of the shrine of Lourdes by night, during the torchlight procession, is a famous illustration (Figure 3): the long and moving ribbon of light unrolls progressively from the grotto to the main esplanade in many circumvolutions. Depending on the number of participants, the ribbon will be long and twisted or the people will gather on the sole esplanade; the shrine will be enlightened in its entirety or partially left in the dark. In the same way, for four days the festival of lights creates a new landscape in the town of Lyon but even more specifically on the 8th of December which is the original celebration day. Then, the inhabitants put candles in their windows, silhouetting the whole city with a ribbon of flickering light. The symbolic aspect of the candle light[4] is very important in these events and almost every shrine town proposes a walk with candles or lanterns; anything but electric light.

Ritual nocturnal practices differ from the nocturnal practices proposed by most tourist authorities by reconnecting with darkness. Banishing electric light for the use of candlelight creates uncertainty; the light could go out at any moment, depending on the weather, the pace of the walk, or the breath of the bearer. It also accentuates contrasts between the luminous and obscure but reduces differences between people at the same time, creating a form of intimacy: the only visible part of a candle bearer is generally her or his face illuminated by the little flame. There is, then, a transforming effect of candlelight, a defamiliarization not only of places but also of people: 'By candlelight, every face is beautiful. Even skin colour contrasts fade away' (Sanctuaires, 2011, p. 86). These words, taken from the brochure of the sanctuaries of Lourdes, were confirmed during the

Figure 3. The torchlight procession in Lourdes in 1964 (This photograph is part of the *Fonds André Cros*, preserved by the city archives of Toulouse and released under CC BY-SA 4.0 license by the deliberation n°27.3 of June 23rd, 2017 of the Town Council of the City of Toulouse).

observation phase of the research. The picture taken at the start of a torchlight procession in Lourdes (Figure 4) shows that, at the beginning of the night and of the procession, the sense of community is strong. Pilgrimage groups arrive one after the other and gather together, each group maintaining a slight separation from the next one. Individual visitors find themselves spread out between the groups. Then, the procession begins. Progressively, as night falls, the gait becomes less self-assured in the dark and rhythms change. Gradually, the groups break up: some are walking at a faster pace, others are looking for someone to rekindle their candle and some are attracted by the powerful light of a flashlight of another visitor, preventing potential obstacles. After about a hundred metres the groups have broken up and become mixed and won't be brought back together before the end of the procession. This is very specific to the night. During the day a lost pilgrim or a lost tourist quickly gets back to her group which she can easily identify thanks to various emblems (for example, a scarf as in Figure 4, or a sign of recognition of the guide).

Figure 4. A group of pilgrims at the beginning of the torchlight procession in Lourdes (M.-H. Chevrier, 2014).

The soundscape by night is as equally important as the landscape. Lourdes and Lyon torchlight processions and the 25th Hour are characterized by canticles, prayers and ringing bells and the *Ave Maria* canticle of Lourdes is a key element. It punctuates the procession, the whole crowd raising their candles at each chorus. In this time of darkness, the visitor gaze, forged by sight *and hearing*, focuses on the essential and a sense of authenticity. Torchlight processions highlight variety in the modes of nocturnal tourist experience: it is still a question of gazing, but also of listening to hymns and prayers and mostly, after a brief learning process, performing those rituals.

More profound emotions both for the pilgrims and secular visitors

In the survey conducted in the shrines, the visitors were asked to give three words to describe the shrine they just visited (Figure 5). Beyond the primacy of spiritual terms such as 'prayer' or 'faith' (39% of the total of the 74 given words), the predominance of 'personal' and 'collective experiences' (25%) was noted above all other 'classical' motives for visiting tourist places. These visits and walks conducted with candles, along with the sacred dimension, play the part of *stimuli* creating emotions *via* senses sharpened by darkness. This temporary 'emotio-spatial hermeneutic' (Davidson & Milligan, 2004) guarantees that the sanctuary is more felt than seen. The place is moving both for pilgrims, in an existential way, and for tourists, in an experiential or experimental way. The response of the visitors to these parameters will eventually generate the desired atmosphere which

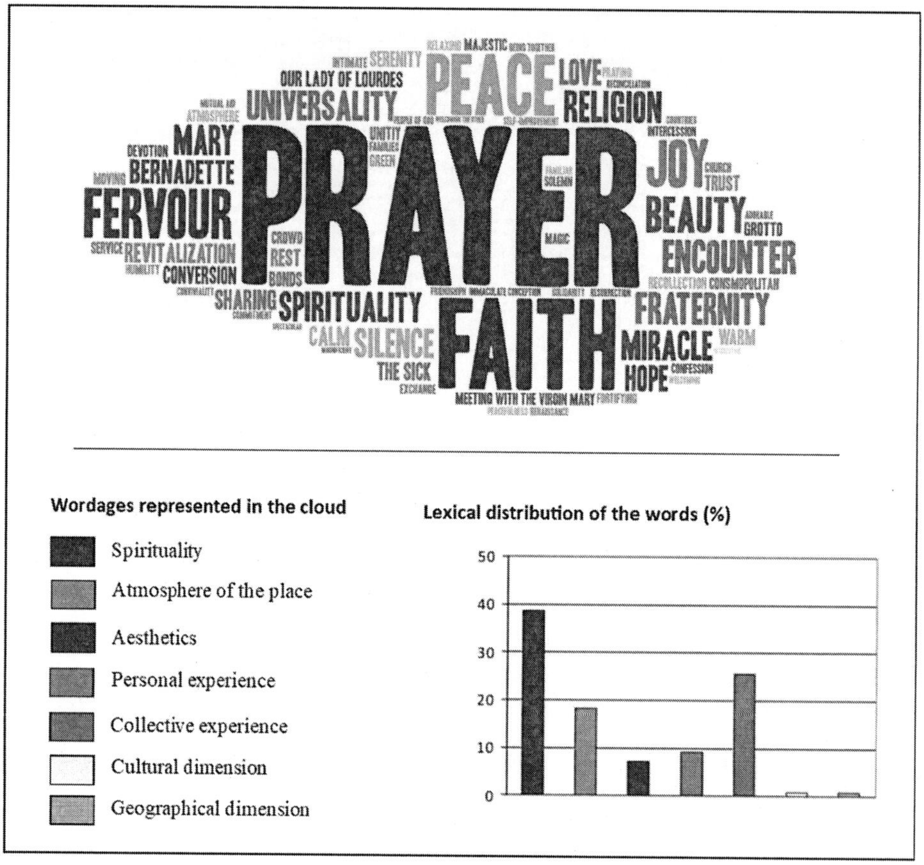

Figure 5. Words given by the visitors of the sanctuaries of Lourdes (M.-H. Chevrier, 2016).

then is a constitutive element of the place, as reminded by the promotional material presented by the tourist office and travel guides of Lourdes:

> 'The atmosphere is one of joy, packed with energy and palpable emotion. It is the most popular moment of Lourdes and the high point of each day. Whether as a participant or a spectator, you cannot help being moved by this « magical » moment which unites men and women from the four corners of the Earth. It is impossible not to be carried away by the power and the beauty of all these lights which easily illuminate the night sky beyond the sanctuary'. (Office de Tourisme de Lourdes, 2016)

The importance of experience is clearly visible in the advertising of the shrines. On the website of Lourdes shrine, the visitor is immediately involved as a 'participant': all the verbs finishing with *-ing*, putting the visitor into action ('visiting the Grotto', 'participating in the Marian torchlight procession', etc.). The same applies for the presentation of the events of the 25th Hour. The visitors become performers during the night.

The second key element of those touristic events is that they are meant to be collective experiences. The short description of the Lourdes procession emphasizes the 'unity' provided by it. Lyon light festival and the 25th Hour share a similar objective: 'Nowadays, the idea is to be together and to stroll through the same streets without clashes and without danger', as stated in 2014 by the Mayor of Lyon's assistant in charge of cultural events, M. Georges Képénékian. Night is often conceived as a time of otherness but the procession of Lourdes begins and ends with invitations to share a few words with nearby participants, first by passing over one's light, then by wishing them a good night. It creates a sense of community, as vividly perceived by the visitors and participants of the procession interviewed (Chevrier, 2016). A young woman interviewed in Lourdes stated:

> 'I was surprised by the Marian vigil. I didn't expect to see such a crowd talking, walk around and pray together.'

Another young participant interviewed added:

> 'In fact, it seems to be a public practice. There are so many people that it is not a private practice, as if we were alone to see the Grotto. So, it actually made me feel like it was a place for sharing.'

Meanwhile, a third stated that:

> 'Relationships with people are totally different: it is like a big family. We approach people we don't know but a bond is immediately forged.'

Likewise, even though most visitors come to the Lyon Festival of Lights with friends or family and stroll the streets in small groups, the Missionaries of the 8th, Christian believers sent by the diocese, go to meet these groups and somewhat break the individualist perspective. This initiative extends to tourists who are also invited to experience a sense of community felt between the pilgrims.

Night, as presented in these examples, is a trial. It supposes sacrifices such as leaving the comfort of one's home, postponing sleep or accepting limited visibility – all these being the opposite of the relaxation sought by most tourists. Conversely to most tourist events, these nocturnal ritual practices are never cancelled and require personal sacrifices. Whatever the weather, the torchlight procession of Lourdes carries on. The visitors persevere and rekindle their candle patiently. The fact that these practices are always maintained, whatever happens, is also of symbolic importance. For instance, despite the cancellation of the Lyon Festival of Lights following the attacks of November 2015 in Paris, the religious procession was maintained, and the people of Lyon put even more candles in their windows as an act of resistance against terrorism. Candlelight was then, more than ever, the metaphor for the victory of light over darkness.

Discussion

Authenticity by refamiliarization and a sense of place

Edensor (2015) underlines that artists seek to defamiliarize places with light artwork. Here, I will argue that nocturnal ritual practices also invite visitors to re-appropriate sacred space. In Lourdes, barriers and lines are put up every evening before the procession to channel the flow of people. This precise itinerary is not one a visitor would usually use so it creates a distinct way of practicing the sacred space (Figure 6). In Lyon and the shrine towns during the 25th Hour, similar measures are taken and the flow of people is controlled in one direction. During the day a visitor can stroll freely through the shrine or the city, but at night there is only one way of going through space. This is partly motivated by security reasons. It reduces risks for the visitors who might get lost or harm themselves stumbling in the dark. These nocturnal itineraries are also thought to give meaning to the place. In Lourdes (Figure 6) the starting point is the Grotto, the heart of the sanctuary and usually the visitor's goal. Starting from the Grotto is a change of perspective. Then, the visitors slowly walk the entire esplanade. By day, this esplanade is only a passage for hurried visitors who may not pay attention to their surroundings. Thanks to the torchlight procession, visitors take their time to go through the entirety of the esplanade and look at the basilicas which are progressively illuminated. This nocturnal procession contributes to the creation of a connection between the place and the visitors who then declare a feeling of being 'at home' in the sanctuary. The Mission of the 8th in Lyon and the 25th Hour of the shrine towns, bear this same idea to re-familiarize the visitors with place and, most specifically, with the sacredness of that place.

Indeed, by perpetrating or reviving ancient traditions, nocturnal ritual practices develop a 'sense of place' (Edensor, 2015). The night events of Lourdes, Lyon and the shrine towns are deeply rooted in history. The torchlight procession of Lourdes, established in 1872, had never been interrupted. Lyon Festival of Lights is a modern version of the initiative of the people of Lyon who, on the 8th of December 1852, put lighted candles at their windows at night, celebrating the inauguration of the statue of the Virgin Mary placed on the rooftop of the Chapel of the Virgin, at the summit of the hill of Fourvière. The events of the 25th Hour also introduce the visitors to the history of the place: most of the towns opt for a narration of tales and legends of the sanctuary (Mont-Saint-Michel for example) and even theatrical reconstructions of historical events or customs (Chartres).

This historicized positioning of the events reinforces the place identity but is sometimes forgotten as it was in Lyon. According to the Mayor of Lyon's assistant in charge of cultural events, M. Georges Képénékian: 'the religious anchorage is still present, but strictly sticking to it or, on the contrary, forgetting it, would be an unfortunate mistake' (G. Képénékian, 2014). However, despite the protests of the City Hall, the creation of

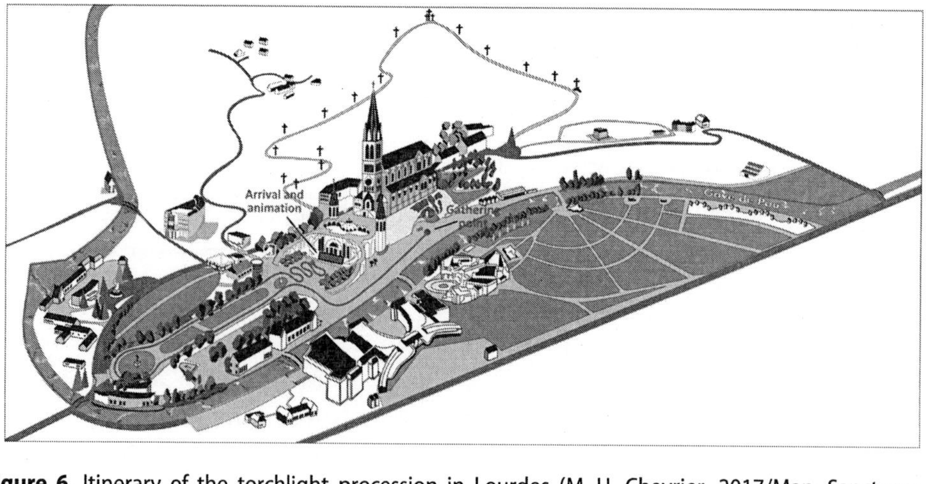

Figure 6. Itinerary of the torchlight procession in Lourdes (M.-H. Chevrier, 2017/Map: Sanctuary of Lourdes)

the Festival of Lights in 1999 was a 'hijacking' of the religious celebration and is still presented as such in the literature (Gravari-Barbas, 2007) and the media, as well as by the Catholic Church. When the Festival of Lights was initiated by the City Hall in 1999, the Church was not involved in the project. It took the diocese of Lyon five years to react and launch, in December 2004, a major event called 'Missionaries of the 8th'. Thus, for the duration of the festival, all churches, usually closed at night, are open until midnight and volunteers are there to welcome visitors with hot chocolate, coffee, music, and diverse types of cultural and religious events. Many of the volunteers wander in the streets all night long. They approach tourists and give them brochures explaining the religious meaning of the feast, and maps registering torchlight processions of the different boroughs and events proposed in the different churches of Lyon. The Church of Lyon set up here an effective tourist strategy, coordinated by the archdiocesan Pastoral of Tourism. This description of the 2009 feast by Etienne Piquet-Gauthier (then chief of the Pastoral of Tourism of Lyon) confirms this

> 'Even if outside it is crowded and cold and there is noise and light, we propose open churches, conducive to dialogue or silence. It is hot, you can sit, rest, drink an herbal tea, write a prayer intention, light a candle at the feet of the statue of the Virgin Mary, pray, talk to a priest ... It is our ambition to make the 8th of December a great diocesan feast during which Christians rise up.' (E. Piquet-Gauthier, 2009)

The festive dimension of the night promoted by the City Hall in a strategy of international competition is obviously present in the 'Mission of the 8th' initiated by the Church (musical animations in and around churches, stall of traditional mulled wine, hot chocolate etc.). But the missionaries also propose, in the midst of artificial lights and loud music, a reconnection with the characteristics of the night much linked to the sacred: silence and darkness. The simplicity of candlelight opposed to the complexity of light art calls for a more authentic experience and a refamiliarization with the place. Reminding visitors of the sacred dimension of these nocturnal practices contributes to deepening the visitors 'sense of place'. The tourist experience, therefore, shifts from recreational or diversionary mode to an experiential, experimental or existential mode.

Using the sacred to create authenticity

Of course, one could object to the previous argument; in the night as discussed here many elements are staged, for instance the religious rituals, or the *son et lumière* in Lyon. Following MacCannell's argument (MacCannell, 1973), the authenticity of these events should thus be questioned. Indeed, the observations, interviews and survey conducted led me to take another look at the idea of tourism developed by MacCannell and at the 'tourist gaze' concept of Urry. Both portray the tourist as a passive visitor, a mere consumer of what is produced or sold by tourist agencies, municipalities and other promoters. This vision has been challenged, notably by geographers (Everett, 2008; Perkins & Thorns, 2001) who highlighted that the visitor is not satisfied with simply contemplation anymore. She needs now to be fully involved, which is specifically demonstrated by the cases studied here. By night, in Lourdes, in Lyon and in the other shrine towns, the visitors (and the inhabitants) are the ones making the landscape. The tourist 'experience' replaces the tourist 'gaze'; the visitor is not a spectator of the event but an actor. If these rituals (which are staged practices) provide the visitors with a sense of authenticity, it is, on the one hand, because of the visitors' own involvement in it while, on the other hand, because these rituals belong to the sacred, which is framed as outside traditional tourism economics. These ritual practices are not organized by the economic agents of tourism but by the ecclesiastical protagonists whose motivations are seen to be more theological than economic, as demonstrated by an agent of Lourdes tourist office during an interview (Chevrier, 2016):

> 'It is true that [in winter] at the sanctuary, supply is reduced. For instance, there is no torchlight procession. But for us, it is the selling point to convince people to stay in Lourdes for a night. So, in winter, we say: 'well ... there isn't'. So we try to convince them [the sanctuary] to organize something all the same: a vigil at the grotto, or anything else for these specific costumers who travel when Lourdes is asleep (Laussu, 2016).'

The attempts of this agent remain unsuccessful for the time being, for the existence of the procession does not respond to costumer's demand but to theological motives. When an event based on a sacred meaning is recreated by municipal authorities or economic agents (as it was for Lyon Festival of Lights), existential authenticity is not prominent among the visitors' feelings. But in the case of Lyon, existential authenticity appeared the moment the Church took its place in the event and sent some believers ('Missionaries of the 8th') to meet the visitors. These missionaries changed the tourist experience and enriched it with an 'empathic understanding' (Wang & Gnoth *in* Jafari & Xiao, 2016). The tourist offices of shrine towns understood it well, creating the '25th Hour' event in close collaboration with the sanctuaries. The feeling of authenticity then intervenes thanks to the sacred dimension of the place and of the practices, coupled with a form of sincerity and trust which seems to erase commercial purposes. 'Cool authenticity' (related to rituals and normative texts) is then validated by the emotional response, individual and collective, and becomes a 'hot authenticity'. Pilgrimage cities and religious rituals satisfy the quest for authenticity of their visitors and performers, particularly through nocturnal rituals and night events.

A more secure experience for all audiences and towns: extending tourist space and time

Pilgrimage cities are popular cities in every sense of the word. Firstly, some attract many international pilgrims and tourists and rank among the most important tourist places in the world (Lourdes, Fatima, Mont-Saint-Michel, etc.). Others are important attractions at a regional or local scale: the sanctuary of Pontmain attracts 200.000 visitors a year, in a French department (Mayenne) otherwise ranked 89 in the tourist ranking, considering the 96 metropolitan departments. Secondly, the survey conducted in the shrines pointed out (Figure 7) that even if the larger part of the visitors are retired people over 60 years old (41.1% of the visitors) who have more free time to go and visit these places, people of all ages are also present in these towns in quite similar proportions, producing an important generational intermingling. Finally, there is also social intermingling because pilgrimage is a popular practice. Christian theology and faith are based on community and brotherhood, implying equality. Pilgrimage is a time to enter a world of *communitas* (Turner & Turner, 1978) where social barriers disappear. So, contrary to many tourist places and tours dedicated to specific social groups, pilgrimage places are places where people mix.

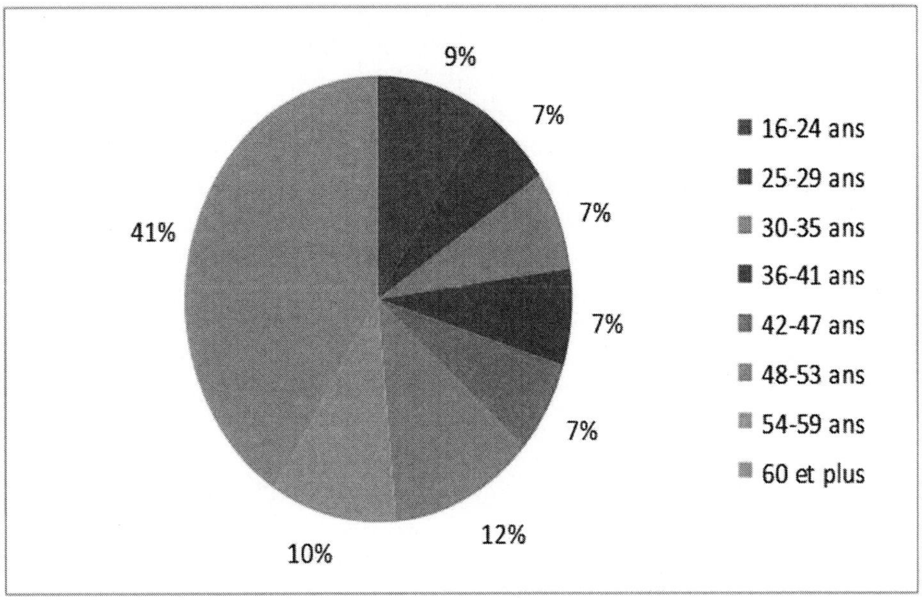

Figure 7. Share of the different visitors age groups in thirteen European sanctuaries (M.-H. Chevrier, 2016)

Nocturnal ritual practices then constitute a climax of popularity and intermingling. According to tourist office and tourist guides of Lourdes, the torchlight procession is the 'most popular event' of the city. Likewise, in Fourvière, during the Festival of lights, according to the President of the Fourvière Foundation managing the place, 'there are loads of people' (Malartre, 2014). The 25th Hour also presents the event as 'an opening', an event 'in which everyone could identify with' (Association des Villes Sanctuaires en France, 2016). In sanctuaries, limitations of age do not apply, and night does not belong only to the young. Nocturnal ritual practices are also adapted for old and disabled people: volunteers are here to help people in wheelchairs or walking with difficulty. The pace of the procession is slow enough for everyone to follow. Thus, ritual nocturnal practices are performed at night by the whole family. Consequently, it is possible to experiment and develop nocturnal family tourism in such places: during the Mission of the 8th in Lyon, for instance, special activities such as living Nativities are dedicated to children. In Lourdes, torchlight processions gather everyone, from little children to elderly people.

Nocturnal ritual practices are then a real selling point for smaller towns which the tourism market often casts aside. Ms Laussu, promotion and communication manager at Lourdes tourist office says that the procession is '[their] selling point to make people stay in Lourdes for one night'. For the tourist office, the torchlight procession represents 'the high point of the day'. This is the 'high point [they] push to the front'. During the pilgrimage season (from April to October), as Ms Laussu (2016) explains:

> You can come in the evening, you can come at night. I am amazed by the number of people who are here, at two o'clock in the morning, at midnight. When there isn't anybody, they come.

Meanwhile, while churches and chapels close at night all around France, Lourdes shrine is never totally closed. The 22 places of worship of the shrine (basilicas, churches and chapels) are closed but the grotto is accessible all night. As Ms Laussu suggested, night is the perfect time for those in search of solitude, peace and silence, far from the festive offer made by capital cities. If early night is festive and illuminated thanks to the procession, profound night is silent and dark. In major cities, silent and dark are the characteristics of non-touristic neighbourhoods: abandoned by public development policies. In pilgrimage cities, the sacred space is planned to value these night attributes and permit the visitors to enjoy it safely *via* adapted lighting and markings. Nocturnal ritual practices extend tourism space-time not only of each day, but also of the year and give it rhythm. The 25th Hour of the shrine towns extends the tourist season until the end of autumn, despite cold temperatures and rain. It is also a symbol for the season's closure. In the Lourdes shrine, for example, the torchlight procession of the 25th Hour is advertised as 'the last procession of the year'. This symbolic aspect added to the unusual proposition of a cultural night event makes it an important promotional operation. Local press as well as national dailies and television present the propositions of the shrines. It certainly creates a dynamic in the shrine towns: each event must be different from the previous one so there is important preparation work. Improvements of shrines and cities, such as the creation of adapted signals and markers, public lighting, etc., are permitted by such events.

Conclusion

Since the 1970s, the development of nocturnal life in large cities has ignored the intrinsic characteristics of the night. Nocturnal spaces are so flooded in light that night became a mere extension of the day. In the current conception and use of the night, there is a desire to escape, even briefly, from daily life. Tourists are in search of this rupture but they can also be on a quest for authenticity. The tourist experience is not only recreational or diversionary, but also experiential, experimental, and even, for pilgrims, existential. These three modes of the tourist experience are rarely compatible with the pleasures currently proposed by much policy concerning the night. However, they are promoted in nocturnal ritual practices. The practices explored above re-enchant the night by preserving and magnifying its characteristics. Highlighting traditional forms of lighting (candlelight) and sound (canticles, bells but also silences), these practices create a mix of uncertainty and intimacy, inviting the visitors to meet other people and re-appropriate space. This authenticity answers to a growing demand of the visitors – whether they are believers or not – and thus represents an important potential for economic development. The success of the Missionaries of the 8th who work on reminding the sacred meaning of Lyon Festival of Lights is a marker of that thirst for authenticity. Based on ancient traditions, these nocturnal ritual practices are not a privilege of capital cities. On the contrary, they are at the heart of the identity and culture of modest towns which have grown thanks to a sanctuary, then have slowly decreased, confronted with the secularization of societies. While the quest for authenticity leads visitors to sacred places once again, the tourist authorities of these modest towns have well measured the importance of developing this offer of night cultural tourism based on ritual practices. Furthermore, it is very much easier and less expensive to develop and it causes less security problems than planning festive events or the creation of a zone dedicated for leisure at night. A real political ambition is now missing for developing this form of night tourism. Capital and regional cities are still trapped in the global competition in which they are engaged and from which authenticity, until now, has been erased.

Notes

1. Religious tourism is defined by Gisbert Rinschede (1992, p. 52) as "that type of tourism whose participants are motivated either in part or exclusively for religious reasons".
2. A canonical procedure of authentication of supernatural events was initiated by the Council of Trent (1545–1563) and clarified in 1978 in an official document of the Congregation for the Doctrine of the Faith.
3. After a last minute cancellation in 2015, the Festival returned in 2016 but with a reduced perimeter, in order to reinforce security faced with terrorism.
4. The flame is a symbol of the victory of light over darkness, as Christ was victorious over death. It also symbolizes the believer's faith as a guiding light in darkness and the fragility of this faith which could alternatively be blown out but also rekindled.

Disclosure statement

No potential conflict of interest was reported by the author.

References

Association des Villes Sanctuaires en France. (2016). La 25ème Heure des Villes Sanctuaires, Samedi 29 October 2016. Retrieved from http://www.visitalencon.com/assets/dossier-de-presse-2016-vsf-v7.pdf

The Association of Shrine Towns. http://www.villes-sanctuaires.com/

Azria, R., & Hervieu-Léger, D. (2010). *Dictionnaire des faits religieux*. Paris: Presses Universitaires de France.

Belhassen, Y., Caton, K., & Stewart, W. P. (2008). The search for authenticity in the pilgrim experience. *Annals of Tourism Research*, *35*(3), 668–689.

Bonte, M. (2017). *Beyrouth, états de fêtes. Géographie des loisirs nocturnes dans une ville post-conflit* (Doctoral dissertation). Available from tel.archives-ouvertes. (tel-01697151).

Chevrier, M.-H. (2016). *Pratiques et valeurs spatiales, pèlerines et touristiques. Grands et petits lieux de pèlerinage aujourd'hui* (Doctoral dissertation). Available from tel.archives-ouvertes. (tel-01449623).

Cohen, E. (1979a). A Phenomenology of tourist experiences. *Sociology*, *13*(2), 179–201.

Cohen, E. (1979b). Rethinking the sociology of tourism. *Annals of Tourism Research*, *6*(1), 18–35.

Daniel, Y. (1996). Tourism dance performances: Authenticity and creativity. *Annals of Tourism Research*, *23*(4), 223–244.

Davidson, J., & Milligan, C. (2004). Embodying emotion sensing space: Introducing emotional geographies. *Social and Cultural Geography*, *5*(4), 523–532.

Edensor, T. (2012). Illuminated atmospheres: Anticipating and reproducing the flow of affective experience in Blackpool. *Environment and Planning D: Society and Space*, *30*, 1103–1122.

Edensor, T. (2013). The gloomy city: Rethinking the relationship between light and dark. *Urban Studies*, *52*(3), 422–438.

Edensor, T. (2014). The rich potentialities of light festivals: defamiliarisation, a sense of place and convivial atmospheres. In J. Meier, U. Hasenöhrl, K. Krause, & M. Pottharst (Eds.), *Urban lighting, light pollution and society* (pp. 85–99). New York: Routledge.

Edensor, T. (2015). Introduction: Sensing and perceiving with light and dark. *The Senses and Society*, *10*(2), 129–137.

Espinasse, C., Gwiazdzinski, L., & Heurgon, E. (2005). *La nuit en question(s): Colloque de Cerisy*. France: Editions de l'Aube.
Evans, G. (2011). Hold back the night: Nuit Blanche and all-night events in capital cities. *Current Issues in Tourism, 15*(1–2), 35–49.
Everett, S. (2008). Beyond the visual gaze? The pursuit of an embodied experience through food tourism. *Tourist Studies, 8*(3), 337–358, London, Sage.
Gravari-Barbas, M. (2007). A la conquête du temps urbain: La ville festive des '24 heures sur 24'. In P. Duhamel & R. Knafou (Eds.), *Mondes urbain du tourisme* (pp. 55–74). Paris: Belin.
Gwiazdzinski, L. (2002). *La nuit dimension oubliée de la ville: entre animation et insécurité* (Doctoral dissertation). Available from tel.archives-ouvertes. (tel-00599146).
Gwiazdzinski, L. (2008). Ambiances nocturnes des villes: Premières relectures en mouvement. Augoyard, J.-F., 1st International Congress in Ambiances, Grenoble, France, Sept. 2008, A La Croisée, 145–152.
Heath, T., & Strickland, R. (1997). The twenty-four hour city concept. In T. Oc, & S. Tiesdell (Eds.), *Safer City Centres. Reviving the public realm* (pp. 170–183). London: Sage.
Jafari, J. (2000). *Encyclopedia of tourism*. London: Routledge.
Jafari, J., & Xiao, H. (2016). *Encyclopedia of tourism*. Cham, Switzerland: Springer.
Képénékian, G. (2014, December 5). Interview by C. Binick and R. Chevassu for *Le Figaro*, Lyon, France.
Koslofsky, C. (2011). *Evening's Empire: A history of the night in early modern Europe*. Cambridge: Cambridge University Press.
Laussu, C. (2016, April 26). Interview by M.-H. Chevrier [audio recording]. President of Atout France's 'Tourism and Spirituality' Cluster, promotion and communication manager at Lourdes Tourist Office. Tourist Office, Lourdes, France.
Lovatt, A., & O'Connor, J. (1995). Cities and the night-time economy. *Planning Practice and Research, 10*(2), 127–133.
MacCannel, D. (1973). Staged authenticity: Arrangements of social space in tourist settings. *American Journal of Sociology, 79*(3), 589–603.
MacCannell, D. (1976). *The tourist. A new theory of the leisure class*. New-York, NY: Schocken Books.
Malartre, G. (2014, April 17). Interview by M.-H. Chevrier [audio recording]. President of Fourvière Foundation. Offices of Fourvière Foundation. Lyon, France.
Ministère des Affaires Etrangères et du Développement Durable. (2014). Pôles d'excellence touristiques. 22 mesures pour faire de la vie nocturne un facteur d'attractivité touristique à l'international. Retrieved from https://www.diplomatie.gouv.fr/IMG/pdf/feuille_de_route_pole_nuit_finalisee_cle4add54-1.pdf
Moufahim, M., & Lichrou, M. (2018). Pilgrimage, consumption and rituals: Spiritual authenticity in a Shia Muslim pilgrimage. *Tourism Management, 70*, 322–332.
Office de Tourisme de Lourdes. (2016). *Lourdes inspirations*. Retrieved from https://www.lourdes-infotourisme.com/web/FR/37-documentation.php
Perkins, H. C., & Thorns, D. C. (2001). Grazing or performing? Reflections on Urry's tourist gaze in the context of contemporary experience in the antipodes. *International Sociology, 16*(2), 185–204.
Piquet-Gauthier, E. (2009). Interviewed by J.-B. Maillard. Chief of the Pastoral of Tourism of Lyon. Lyon, France.
Rinschede, G. (1992). Forms of religious tourism. *Annals of Tourism Research, 19*, 51–67.
Roberts, M., & Eldridge, A. (2009). *Planning the night-time city*. London: Routledge.
Sanctuaires Notre-Dame de Lourdes. (2011). *Expliquez-moi les processions* [Explain processions to me], Lourdes, NDL Editions.
Selwyn, T. (1996). *The tourist Image: Myths and myth making in tourism*. Chichester: Wiley.
Turner, V., & Turner, E. (1978). *Image and pilgrimage in Christian culture: Anthropological perspectives*. Oxford: Basil Blackwell.
Urry, J. (1990). *The Tourist Gaze: Leisure and Travel in Contemporary Studies*. London and Newbury Park: Sage Publications.
Wang, N. (1999). Rethinking authenticity in tourism experience. *Annals of Tourism Research, 26*(2), 349–370.

Residents versus visitors at light festivals in cities: the case of Barcelona

Raquel Camprubí and Lluís Coromina

ABSTRACT

In recent decades, cities around the world have created light events in order to improve their competitive position and make themselves more attractive to tourists. The relative novelty of these events means that the academic literature has not yet developed a corpus of knowledge outlining the visitor profile and behaviour characteristics. Thus, a study of the Barcelona event 'Llum BCN' was conducted in order to characterise the demand and detect possible differences in demand behaviour according to whether the attendees are residents or visitors. Comparisons between residents and visitors show that residents are more highly motivated to attend the event than visitors. On the other hand, residents show higher levels of satisfaction in variables related to the type of illumination during the event, particularly innovative design elements.

Introduction

In recent years, a significant number of cities have begun to use illumination in ways beyond those which are purely functional. Illumination of public spaces such as streets, squares and gardens for safety and commercial purposes has evolved, resulting in a new way of understanding lighting management (Mantei, 2012). City planners and marketers have realised that lighting is a powerful tool to generate new tourist attractions, or highlight existing ones (Demers, 2010). 'The ways in which light transforms space is complex and multiple' (Edensor, 2012, p. 1106); it enhances the beauty of monuments and other tourist nodes, and can be used to differentiate day-time spaces from night-time spaces or create a nocturnal tourism image based on specific city icons.

In general, there are three different types of illumination: (a) permanent illumination; (b) seasonal illumination; and (c) ephemeral illumination (Mantei, 2012). Although the two first types are common in all cities, ephemeral illumination has been postulated as being innovative in its ability to increase the attractiveness of a tourist site over a short period of time. In general, cultural events have been developed as a way of improving a city's image, adding life to its streets and offering residents a renewed pride in their own city (Richards & Wilson, 2004). These events also promote and stimulate tourism development (Quinn, 2005), and 'an increasing awareness of the potential of lighting festivals as touristic attraction and for city marketing' (Giordano & Ong, 2017, p. 705) has emerged. These are the main reasons why cities worldwide have created events based on spectacles of light, colour and sound. Technical advances have led to new techniques such as projecting images onto monuments and historical buildings (Edensor, 2012). Lighting is also used in other, ever more creative ways.

As this is a relatively new phenomenon until now there have been very few studies in the academic literature (Edensor, 2012; Giordano & Ong, 2017) focusing on the topic of illumination as a tourist attraction, or its impact on tourist demand. Thus, the aim of this paper is to analyse the profile and behaviour patterns of attendees at a light event, taking into consideration possible differences arising from whether they are residents or visitors. In this study, attendees are considered individuals who go to see the event, and can be classified as residents (people who live in the city), or visitors (non-residents). 'Llum BCN', an annual light event held in the city of Barcelona (Spain), was taken as a case study.

Literature review

Lighting in urban spaces

Since electricity was first invented, night-time illumination of cities has become a critical issue. Initially, the function of lighting was to ensure safety and order (Caralt, 2009), but it was gradually employed for other purposes.

During the 1990s, 'mise en lumière' (the illumination of monuments, historical buildings and public spaces) became widespread in numerous cities and small villages in France, and was considered part of the development of urban spaces (Bourgeois, 2002). Technological advances have helped this phenomenon to spread to other places in recent decades. However, there is evidence it was used before this time, such as the illumination of Concordia Square in Paris in 1944, or the illumination of the main Parisian monuments in 1937 (Bourgeois, 2002). Illuminating urban attractions has now spread to other cities around the world.

According to Demers (2010), a city can reinvent itself at night by contrasting two opposing qualities: darkness and light, thus, lighting has become an essential element of urban landscapes. Illumination enhances and beautifies monuments and other spaces, differentiating day-time spaces from night-time ones, and ranking monuments and other spaces in order of importance. Using light is a way of giving value to urban spaces by creating an image of universal value imposed on the quotidian idea of the city (Bourgeois, 2002). This offers a more positive and solid image of the city, particularly tourism spaces.

Up to this point, we have focused implicitly on permanent lighting, but according to Mantei (2012), seasonal illumination such as Christmas decoration in public spaces; or ephemeral illumination, where light is the main attraction of the spectacle, is also relevant. The academic literature (Bourgeois, 2002; Cousseau, 2015) has particularly centred on the first two types of illumination (permanent and seasonal), as these are considered more common in cities, and essential in urban spaces. From an innovation perspective, this article contributes to the academic literature by exploring ephemeral lighting, and the impact of the increasing popularity of illumination in festivals and events worldwide.

Lighting events in urban spaces

Urban spaces need to be able to survive in a growing, competitive environment. Cities have found a way to increase their competitiveness through lighting events, thus creating new tourist attractions (González Reverté, 2010; McHone & Rungeling, 1999), while simultaneously increasing leisure activities for residents (González Reverté, 2010).

Events have different purposes, and not all the events held in a city are directed towards tourism. Many have other objectives (Getz, 2008) such as satisfying residents (González Reverté, 2010), or stimulating a particular commercial sector. Getz and Page (2016) outline some of the main benefits a city derives from holding events, and state that apart from creating an attraction for tourists, events can also do the following: create a positive tourist image of the place; contribute to creating or repositioning a brand; fight the seasonality of tourism demand; be a catalyst to catch attention of multiple market segments; and strengthen local identity, enhancing community-building. Additionally, Richards and Wilson (2004) also point out the potential benefits and relatively low cost of events in differentiating places and attracting interest from the mass media.

According to Getz and Page (2016), various types of events exist, depending on their dimensions, periodicity, and specialisation. Among these, cultural events have received special attention in the academic literature. Cultural events can be categorised into five types: historical-cultural festivals; art festivals; carnivals; religious events; and other types of cultural events such as concerts or theatre (Getz, 2008; Kay, 2004). In this context, light festivals can be categorised as art festivals or religious events, depending on the place they are held and their origins, and they can also be considered as late-night cultural events (Evans, 2012; Giordano & Ong, 2017).

In the context of cultural events, light plays an important role in multiple festivities. Examples are the Bala Chaturdashi Festival in Nepal, Saint Lucy's Day in Scandinavia, or Dussehra in India. Contemporary light festivals have emerged as a way to add to experiences cities already offer and 'have been perceived by local authorities [...] as a valid alternative to the traditional night-time economy based on drinking and [...] as an attraction capable to extend temporally and geographically the distribution of tourism activity' (Giordano & Ong, 2017, p. 705). Light festivals have been postulated as combining enhanced cultural heritage and artistic staging, and the illumination of nodes is frequently accompanied with sound and/or artistic works. For Demers (2010) light events transform the night within a special time-span, and aim to build bridges between individuals and the 'other' city. This means using darkness as the protagonist of the nocturnal city, and using light to show various every-day features of the city. However, several authors have given advice about potential homogenisation (Quinn, 2005), and the risk of suffering from 'serial reproduction' or 'serial monotony' (Richards & Wilson, 2004) in festivals. In this context, cities that have created light festivals also have to face the risk of homogenisation and trivialisation of lighting, due to the high proliferation of using these types of events as an urban development strategy (Mantei, 2012).

Although this type of illumination appears to be a recent phenomenon, in fact, it originates from the end of the nineteenth century and the beginning of the twentieth century, when fountains were lit up (Caralt, 2009). Undoubtedly, Lyon's 'Fête de la Lumière' is the pioneer of light festivals as it featured many of the technical advances or artistic creations for the first time. Originally, 'Fête de la Lumière' was a religious festival, which later grew into a secular event with 80 artistic projects and drawing 800,000 visitors. According to Giordano and Ong (2017), Lyon (France) was the first city to fully adopt the new lighting principals developed in 1980, thus becoming an urban laboratory for light design methodologies and tools. Lyon has therefore inspired other cities around the world to develop their own light festivals (Melbourne, London, Jerusalem, Baltimore, Barcelona, etc.), converting these types of events into a global phenomenon. The majority of these festivals are created ex-novo, based on night-time animations combining light and sound, artistic performances, or images projected on the walls of particular buildings. Numerous examples of successful experiences can be found: the Bunge grain silos in Quebec were a set of unknown buildings which were recognised as urban monuments and brought to the public eye with the 'Moulin à images' event (Demers, 2010); 'White Night Melbourne' attracts more than 550,000 visitors after 5 years (Woodhead, 2017); and 'Blackpool Illuminations', held each autumn, runs for sixty-six days and includes more than 500 scenic designs (Edensor, 2012).

Light festivals stem from the point where cultural and tourism policies converge, where decision-makers try to balance the benefits derived from the ephemeral illumination, not only for tourists but also for the local community. In this context, Mantei (2012) points out that lighting festivals enable a city's inhabitants to reappropriate their city, for instance, through revitalising local cultural traditions, creating and reinforcing a dynamic and festive dimension, or emphasising the need to restore heritage. Nevertheless, it seems harsh to criticise the real role and purpose of festivals in urban spaces (Quinn, 2005, p. 29) when we take into consideration that festivals sometimes have an 'undoubted potential in animating communities, celebrating diversity and improving quality of life'. Quinn's contributions (2005, 2010) highlight the difficulties decision-makers face in order to combine cultural and tourism policies successfully.

The attendees at events

Events have gained importance when it comes to improving the competitiveness of tourism destinations, particularly cities. In essence, cultural events are characterised as ephemeral, and require spectators' attention to give meaning to their existence. That is to say, the events need visitors to participate, watch, learn and enjoy them (Walsh-Heron & Stevens, 1990).

In this context, McHone and Rungeling (1999) indicate that in order to carry out a tourism strategy based on diversification, it is important to know the demographic characteristics of the new visitors, as well as their behaviour patterns and expected expenditure. Following on from this research, various other studies have emerged that attempt to reveal the profile of visitors at events. For example, Nicholson and Pearce (2000), apart from characterising the attendees at four Australian cultural events, also focused on attendees' behaviour, and included variables related to the composition of the travel group or differentiated between first-time visitors and repeat visitors in their study. Light (1996) carried out other relevant research in this field, emphasising the characteristics of visitors at Caerphilly Castle (Wales), and comparing differences in visitor profiles during periods when the event took place and periods when there was no event. This study researched the visitors' profile and variables related to visitor behaviour during the visit, as well as the level of satisfaction derived from the visit. Light (1996) saw no significant differences in visitor profiles whether there was an event or not.

A number of studies focus more specifically on the motivations that lead people to a particular event (Backman, Backman, Uysal, & Sunshine, 1995; Crompton & McKay, 1997; Kim, Uysal, & Chen, 2001; Scott, 1996; Uysal, Gahan, & Martin, 1993). These studies show various factors which motivate people to attend the event, and these have been adapted to the characteristics of the event analysed in this case study. For example, Crompton and McKay (1997) established a total of 28 motivational aspects, aggregated in six factors. In parallel to these studies, other studies on visitor behaviour at cultural events also evaluated visitor satisfaction. For example, Light (1996) demonstrated that, in general, a visit to a heritage site during the event brings a higher level of satisfaction.

Another area that researchers have centred on is whether those attending an event are residents or visitors. In this context, Stevens and Shin (2014) point out that attendees at festivals try to find out the meaning of a place as a physical reality, however, this reality is not necessarily alienated from individual citizens. This suggests that the behaviour of residents during a festival is different from the behaviour that they have on any other day in the same place; so, in some aspects, residents change their behaviour to be more like 'visitors'. For instance, Snelgrove, Taks, Chalip, and Green (2008) concluded that no differences in characteristics or behaviour were found between residents and visitors who attended a sports event. In contrast, Herrero, Sanz, Bedate, and Barrio (2012) found significant differences in the level of expenditure between tourists and residents participating in a music festival, identifying that tourists spend more than locals. Along the same lines, McDowall (2010) concluded that residents and visitors who attended The Tenth Month Merit-Making Festival (Thailand) differed to a certain degree in several areas: festival information sources accessed, motivations for attending the festival, rating of the performance, and overall satisfaction with the festival.

Although the literature centres on analysing tourist demand for various types of cultural events, there are no specific in-depth studies analysing consumer behaviour of attendees in light festivals. Thus, this paper fills this gap in the academic literature and contributes to the knowledge on attendees, specially differentiating residents and visitors to this type of event.

Methodology

This study focuses on the 'Llum BCN' festival, held annually in Barcelona (Spain) since 2012 during the Santa Eulàlia festivities in mid-February. The city's old quarter is filled with a variety of art installations over three days, during which night and light are partnered in a light show which illuminates and colours the city's landmarks. The show seeks to interact with the public by linking stories, technology and heritage together. The art works on display during the festival are created by teachers and students from the city's schools of art, interior design, architecture and lighting. Renowned artists such as Michela Mezzavilla, Antoni Arola, Martín Azúa and Jordi Pont have participated in the festival with their artistic designs. In 2016, 'Llum BCN' reached its fifth edition and included 24 art installations and 3 itineraries around the Gothic quarter (Figure 1). This is one of the most frequented quarters in the city (Gutiérrez, García-Palomares, Romanillos,

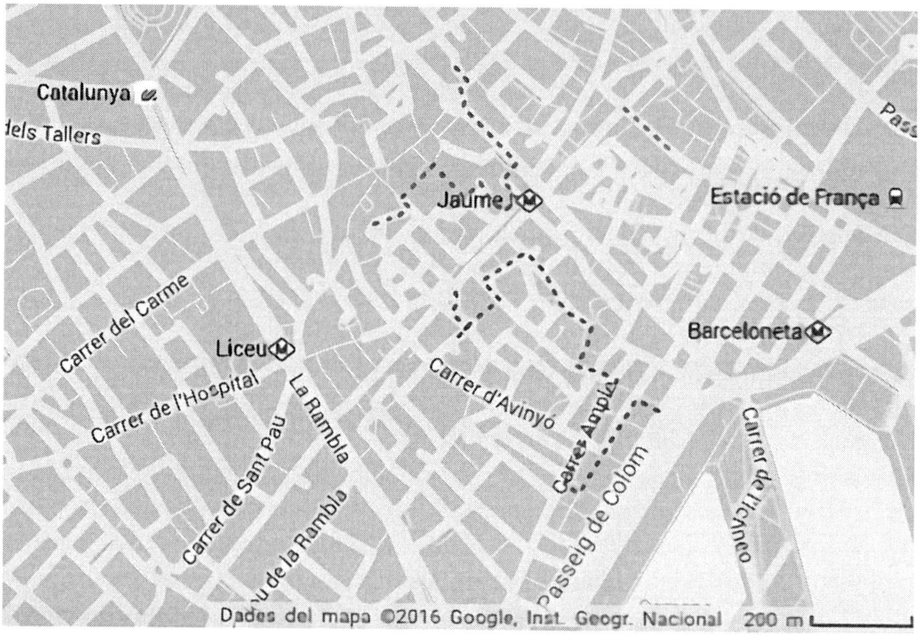

Figure 1. Itineraries for 'LlumBCN'.
Source: Barcelona City Council.

& Salas-Olmedo, 2017) due to the high concentration of tourism attractions and heritage sites.

The original aim of the light festival in Barcelona was to introduce residents and visitors to lighting as a growing artistic art form by lighting up the city's landmarks and, at the same time, give residents the opportunity to rediscover the city. In addition, this was also a way for the City Council to pay tribute to the artist Carles Buïgas and the Monjuïc light fountains inaugurated in 1929 during the Universal Exposition. The first edition was based on lighting unique monumental buildings in the city such as the Sagrada Família, the Arc de Triomf or Casa Batlló but over the years, specific itineraries have been drawn up that also include art installations in patios, squares and streets around the old quarter. Since it first began, 'Llum BCN' has evolved from a show illuminating the city's landmarks in a simple way to a display of real artistic expression and creativity. Visitor numbers have increased steadily year after year, rising to an estimated 170,000 visitors at the last edition. It is important to highlight, however, that attracting residents is the main target of the event.

This event is organised by the City Council of Barcelona. However, the involvement of teachers and students from the city's schools of art, interior design, architecture and lighting give the event a more participatory dimension and contributes to promoting local, talented artists, students and teachers.

Data collection was carried out by two interviewers using tablets. The questionnaire included questions regarding the attendees' profile, motivations, sources of information, recommendations, and satisfaction with the event. It was conducted between 7.00 and 11.00 pm each day of the event. The festival has two itineraries, and interviewers used more than one data collection point on each itinerary in order to randomise the sample. A proportional sample was collected each day of the festival. Some cases were eliminated due to inconsistencies, and the final sample was a total of 182 valid cases with a sampling error of 7.25%. For the data analysis, a significance level of 5% indicates a 5% risk of concluding that a difference exists when, in fact, there is no difference. This indicates a confidence level of 95%, which means there is a 95% probability a difference exists when there is an actual difference.

Data analyses were carried out using univariate and bivariate statistics in order to study consumer behaviour in illumination events. Contingency tables have been specifically used, with the statistical χ^2 test and Cramer's V, in the case of pairs of qualitative variables. Likewise, mean differences and analysis of variance (ANOVA) tests were used to analyse the relationship between numerical variables (for instance, motivations) and qualitative variables (whether the attendee is resident or visitor).

Results

This section presents the results related to attendees of the 'Llum BCN' event. Firstly, the results of the socio-demographic profile of the attendees at the event are presented. Secondly, behavioural variables such as previous visits, motivations, levels of satisfaction, sources of information and recommendations are explained. Finally, differences between residents (who live in Barcelona city) and visitors are analysed in order to explain attendees' behaviour.

In relation to the socio-demographic variables, Table 1 shows that proportionally more females (64.3%) than males (35.7%), visit 'Llum BCN'. The average age is 36.6 years (standard deviation = 12.17 years). The results, therefore, show fairly young attendees, considering that half of them (50.6%) were under 35 years.

The level of education is mainly university (81%), followed by secondary education (18.3%). In general, most attendees are employed (61.3%). It is also worth highlighting that a significant number of attendees at the event were students (19.9%). Concerning the origin of attendees, 52.2% are residents, while 47.8% are visitors. Foreign visitors mainly come from France and the United Kingdom. Finally, the majority of visitors are accompanied by family members (34.3%) or a partner/spouse (30.9%).

Table 1. Attendee's profile.

	Frequency	%		Frequency	%
Gender			*Employment status*		
Male	65	35.7	Self-employed	20	11.0
Female	117	64.3	Employed	111	61.3
			Student	36	19.9
Age			Other	14	7.7
Under 35	92	50.6			
35–44	40	22.0	*Place of residence*		
45–54	33	18.1	Barcelona	95	52.2
55–64	11	6.0	Outside Barcelona	87	47.8
Over 64	6	3.3			
			Companion		
Education			Family	61	34.3
Primary	1	0.6	Partner / Spouse	55	30.9
Secondary	32	18.3	Friends	34	18.0
University	142	81.1	Alone	28	15.7

Table 2. Main reason.

	Total (%)	Visitors (%)	Residents (%)
Not the main reason	79.4	92.0	67.0
The main reason	20.6	8.0	33.0
Total	100	100	100

Note: Cramer's $V = 0.308$; p-value = .000.

Tables 2, 3 and 4 present data on the following: whether the festival was the main reason for the visit; previous knowledge of the festival; and previous visits to the event. Table 2 shows that for around 20% of total attendees (residents and visitors), the festival was the main motivation for attending, and that there is a significant difference (Cramer's $V = 0.308$; p-value = .000) between residents and visitors. The festival as the main reason for visiting is especially low (8%) for visitors. These results confirm that this event is directed at residents, and for the majority of visitors, is complementary to their visit to Barcelona. Table 3 shows that almost half of the total attendees had prior knowledge of the festival (46.2%), especially residents (66.3%); differences between residents and visitors are also statistically significant (Cramer's $V = 0.423$; p-value = .000). As expected, residents have more prior knowledge of the festival compared to visitors. Table 4 shows that 39% of the total attendees are repeating the visit to the event, with a higher level of repetition from residents (54.7%) compared to visitors (21.8%), the difference being statistically significant (Cramer's $V = 0.337$; p-value = .000).

Specific motivations are determining factors and characterise demand for events. In order to analyse motivations for visiting the festival, attendees were presented with a total of 12 statements. These statements are rated using a 5-point Likert scale, ranging from 'totally disagree' (score 1) to 'totally agree' (score 5). Table 5 shows the average for each motivation ('Total' column), the main motivations to attend the festival is 'to see new and different things', 'to enjoy a special event' and 'to search for novelty'. In summary, the main motivations are those statements highlighting factors related to uniqueness and innovation.

In order to study whether differences in motivations exist between residents and visitors, Table 5 shows average values and the statistical significance for difference in averages. The results show that residents have higher levels of motivation for attending the event than visitors. Both groups give 'to see new and different things' the highest value, but

Table 3. Previous knowledge of the event.

	Total (%)	Visitors (%)	Residents (%)
Not previous knowledge	53.8	75.9	33.7
Previous knowledge	46.2	24.1	66.3
Total	100	100	100

Note: Cramer's $V = 0.423$; p-value = .000.

Table 4. Previous visits.

	Total (%)	Visitors (%)	Residents (%)
First visitor	61.0	78.2	45.3
Repeater	39.0	21.8	54.7
Total	100	100	100

Note: Cramer's $V = 0.337$; p-value = .000.

Table 5. Attendee's motivations.

Motivations	Total	Visitors	Residents	*p*-Value
To see new and different things	4.08	3.63	4.48	***
To enjoy a special event	3.99	3.61	4.33	***
To search for novelty	3.98	3.51	4.41	***
To enjoy the festival mood	3.83	3.42	4.20	***
To enjoy a unique environment	3.83	3.63	4.01	n.s.
To escape from the routines of life	3.81	3.41	4.17	***
To enjoy culture in its cultural / historical environment	3.78	3.41	4.12	***
To see the event with my friends and family	3.70	3.38	3.99	***
To increase my cultural knowledge	3.59	3.26	3.89	***
To relieve daily stress	3.28	2.94	3.58	***
To be with people who enjoy the same things as I do	3.22	2.95	3.46	***
I hope the benefits will satisfy my personal needs	3.16	2.78	3.51	***

Note: *** p-value < .05; n.s. = statistically non-significant differences.

differ in opinion regarding the second highest value. Here the residents chose 'search for novelty'. For visitors, the main motivation is also 'to enjoy a unique environment' (average = 3.63), followed by 'to enjoy a special event'.

Information sources

One of the fundamental tools for individuals' decision making when organising a visit is information sources (Coromina & Camprubí, 2016; Gitelson & Crompton, 1983; Li, Pan, Zhang, & Smith, 2009; Llodrà-Riera, Martínez-Ruiz, Jiménez-Zarco, & Izquierdo-Yusta, 2015). Several authors have studied how information sources are used in tourism from different perspectives (Snelgrove et al., 2008). In order to ascertain the attendee's perspective, they were asked which sources of information they used to find out about the festival. The intensity of use for each information source was also reported using a 5-point Likert scale which ranged from 'little used' (1) to 'widely used' (5).

Table 6 shows that the sources of information most used are recommendations from family and friends (38.4%) and the official website (31.3%). Regarding the average intensity of use for each of these information sources, Table 6 shows that recommendations from family and friends are also the most useful when deciding to visit the festival, followed by tourist brochures and the official website. It should be noted that those attendees who used tourism blogs to search for information about the event (5.2%), point out that this information source is not very useful (1.78 on a 5-point scale), and is the least used of the various information sources. Differences between residents and visitors regarding the average intensity of the sources have been tested, and statistically non-significant differences were found. Thus, the average intensity of the information sources is the same whether the attendees are residents or visitors.

Table 6. Information sources used and level of utilisation.

	% of use	Average intensity of use
Recommendation from family and friends	38.4	4.28
Official website	31.3	3.67
Tourism brochures	24.3	3.85
Internet search (Google etc.)	16.6	3.43
Press, radio and television	16.6	3.43
Social network sites	13.6	3.61
Tourism blogs	5.2	1.78

Table 7. Satisfaction with the festival.

	Total	Visitors	Residents	*p*-Value
I am pleased with my decision of visiting the festival	4.15	4.18	4.13	n.s.
The festival satisfies my expectations	3.99	4.10	3.90	n.s
The courtyards (patios) decoration is innovative	3.85	3.89	3.82	n.s
I am pleased with the illumination of the event and the monuments	3.80	4.05	3.61	***
The lighting shows are innovative	3.69	3.89	3.54	***
The festival is well organised	3.64	3.66	3.63	n.s
The festival is not crowded	3.39	3.44	3.36	n.s

Note: *** *p*-value < .05; n.s. = statistically non-significant differences.

Satisfaction, recommendations and future behaviour

One of the most important aims of evaluating attendees behaviour is to determine the level of satisfaction of the experience as this shows the extent to which attendees expectations have been fulfilled. Additionally, it is also possible to determine behaviour such as possible repeat visits, or recommendations given to others regarding a place or event visited.

In order to assess the level of attendee satisfaction with the festival, seven statements were given (Table 7). Each statement had a response ranging from 1 to 5 on a 5-point Likert scale, where 1 means 'poorly satisfied' and 5 means 'very satisfied'. The results confirm that the festival offers a satisfactory experience for its attendees. Table 7 shows that the attendee is particularly satisfied with their decision to visit the festival (4.15) and the fulfilment of their expectations (3.99). They also assess aspects related to illumination and innovation positively. This is followed by the lighting shows and the organisation of the event itself. It should be noted that a minimum score is given to overcrowding at the festival (3.39).

Statistically significant differences were found in two statements when mean differences in satisfaction between residents and visitors were compared. Visitors have a higher level of satisfaction than residents for two statements related to illumination: 'pleased with the illumination of the event and the monuments'; and 'the lighting shows are innovative'. There are statistically non-significant differences between residents and visitors for the rest of the satisfaction statements.

Other factors related to attendee satisfaction and post-event behaviour are willingness to repeat the visit in the future, and willingness to recommend the event. Firstly, in relation to the willingness to repeat, the study found that although there is a high degree of satisfaction with the festival experience, more than half of the attendees do not show willingness to return to the next edition of the festival (52.5%). 41.3% declare their intention to repeat, and only 6.1% confirm their intention to not return to the event in the future (Table 8). Table 8 also tests the statistically significant relationship between the type of attendee (with two categories: 'visitor', and 'resident') and the intention to repeat

Table 8. Future repetition to the festival.

	Total (%)	Visitors (%)	Residents (%)
Yes	41.3	27.6	54.3
No	6.1	9.2	3.3
I don't know	52.5	63.2	42.4
Total	100	100	100

Note: Cramer's $V = 0.280$; *p*-value: .001.

(with the categories 'Yes', 'No', 'I don't know'). A statistically significant relationship was found between residents and visitors, (Cramer's $V = 0.280$, p-value: .001). 54% of residents affirm their willingness to return to the next edition, while 63.2% of visitors state that they do not know if they will visit again or not. The high percentage of visitors stating that a future visit is doubtful attribute this to the distance between Barcelona and their place of residence, rather than factors related to dissatisfaction with the experience.

Secondly, an 11-point Likert scale ranging from 'unlikely' (0) to 'very likely' (10) was used to evaluate willingness to recommend the event to family and friends. The average score for recommending the festival is 6.85 points. In particular, when asked about the probability of recommending the event, 70% of the attendees gave a value of between 6 and 8 points. No significant differences were found when comparing the average recommendation between residents (7.0), and visitors (6.7), ($F = 1.723$; p-value = .191). Therefore, there are no differences between attendees, whether residents or visitors, when recommending the event to family and friends.

Conclusions

Light festivals have become a strategic tool used by cities worldwide to increase their attractiveness and generate a positive image (Demers, 2010). However, previous research on light festivals fails to focus sufficiently on attendees' profile and behaviour. Thus, this study fills a gap in the literature by exploring differences between residents and visitors.

This paper contributes to the academic literature by providing a better understanding of the profile and behavioural patterns of attendees to a light festival. In the case of the light festival 'LLumBCN', findings illustrate that attendees are mainly women, aged under 35 years, with university studies and are employed. These results are consistent with previous research (Herrero et al., 2012; Light, 1996) in cultural events literature.

The findings also prove substantial differences in the behaviour of residents and visitors, along the same lines as the study carried out by Herrero et al. (2012). For instance, the results show that the fact that attendees try 'to see new and different things' is a relevant factor in encouraging people to visit this type of event. For residents, in particular, going to the light festival shows they enjoy the city from a different perspective. This adds value to the city, and provides the opportunity for them to discover it in an especially charming way. This is shown by a higher score in all the motivational aspects assessed. Along these lines, Snelgrove et al. (2008) also proved that motivations to visit sports events varied whether the attendee was a resident or a visitor, and McHone and Rungeling (1999) detected different motivations between intrastate and interstate visitors to an exhibition. In parallel, previous experience of prior editions of the light festival, or previous knowledge of the event, are also relevant factors that influence people and encourage them to visit the festival, especially in the case of residents. In this regard, Light (1996) also pointed out the relevance of previous visits as an incentive to repeat visits to Caerphilly Castle during event days. In relation to information sources, for both types of attendees, the most important source was recommendations from friends and family, followed by the official website, and tourism brochures. These results coincide with a study carried out by Shanka and Taylor (2004), which highlighted word-of-mouth as the most popular information source. Concerning satisfaction, findings reveal that the level of satisfaction is high for both residents and visitors. This will lead to future behaviours which may contribute to the success of the event. In particular, results point to higher satisfaction amongst visitors; particularly, on variables related to the type of illumination during the event, highlighting its innovative component. These results are consistent with Herrero et al. (2012) who also detected higher satisfaction for visitors. Regarding the desire to attend a new edition of the festival, this may depend on the origin of the visitor: future visits will be more likely for residents than visitors. Finally, the tendency to recommend the light festival to others is the same for residents and visitors.

Therefore, 'Llum BCN' is postulated as another, alternative way to consume the city in spaces where artistic spectacles of light and sound are located, which is mainly the old quarter. This type of event enables residents to discover different facets of the city, while visitors see it is an extra motivation to visit the city. From the point of view of cultural policies of the city, 'the role of festivals gradually broadened to assume social dimensions' (Quinn, 2010, p. 268). In this case, the city's cultural policy is important, because this event strengthens the local community through the involvement and active participation of teachers and students from the city's schools of art, interior design, architecture and lighting, who help organise the event. Nevertheless, these kinds of events have to face difficulties resulting from having to make decisions on both cultural and tourism policies (Quinn, 2005, 2010), as well as the risk of homogenisation and trivialisation (Mantei, 2012). In the particular case of Barcelona, the high tourism density where the light festival is held can also be an opportunity for residents to reappropriate the city, and to put pressure on a neighbourhood where over-tourism is already present.

To conclude, this study adds to the academic literature by giving deeper insight and consistent results in the area of events, particularly the underexplored area of light festivals as an event type. Findings showed some consistency with previous research conducted in this field. However, more research is needed in order to determine whether attendees at this type of festival differ from those who visit other types of cultural events and festivals. Along these lines, future research could replicate this study at the same festival using a larger sample, or undertake a study at other light festivals in order to determine if the attendees' profile and behaviour share similar patterns to this event. This study, together with upcoming research, can help practitioners develop a light festival marketing-mix which is more appropriate to their needs.

Disclosure statement

No potential conflict of interest was reported by the authors.

ORCID

Raquel Camprubí http://orcid.org/0000-0001-7084-7889
Lluís Coromina http://orcid.org/0000-0003-0769-0633

References

Backman, K. F., Backman, S. J., Uysal, M., & Sunshine, K. M. (1995). Event tourism: An examination of motivations and activities. *Festival Management and Event Tourism*, *3*, 15–24.

Bourgeois, J. (2002). Le monument et sa mise en lumière. *L'homme et La Société*, *145*, 29–49.

Caralt, D. (2009). *Espectacles de masses a Barcelona (1929).* Carles Buïgas i l' art de l ' aiguallum. In XI Congrés d'Història de Barcelona – La ciutat en xarxa, pp. 1–25.

Coromina, L., & Camprubí, R. (2016). Analysis of tourism information sources using a Mokken scale perspective. *Tourism Management, 56*, 75–84. doi:10.1016/j.tourman.2016.03.025

Cousseau, A. (2015). La iluminación de las ciudades y su imagen nocturna, caso de Lyon y Barcelona. *Papers de Turisme, 57*, 28–46.

Crompton, J. L., & McKay, S. L. (1997). Motives of visitors attending festival events. *Annals of Tourism Research, 24*(2), 425–439.

Demers, A. (2010). L'Offre nocturne de la ville de Québec: vers un tourisme de la nuit. *Rabaska: Revue D'ethnologie de L'Amérique Française, 8*, 43–19. doi:10.7202/045253ar

Edensor, T. (2012). Illuminated atmospheres: Anticipating and reproducing the flow of affective experience in Blackpool. *Environment and Planning D: Society and Space, 30*(6), 1103–1122.

Evans, G. (2012). Hold back the night: Nuit Blanche and all-night events in capital cities. *Current Issues in Tourism, 15*(1–2), 35–49.

Getz, D. (2008). Event tourism: Definition, evolution, and research. *Tourism Management, 29*(3), 403–428. doi:10.1016/j.tourman.2007.07.017

Getz, D., & Page, S. J. (2016). Progress and prospects for event tourism research. *Tourism Management, 52*, 593–631. doi:10.1016/j.tourman.2015.03.007

Giordano, E., & Ong, C. E. (2017). Light festivals, policy mobilities and urban tourism. *Tourism Geographies, 19*(5), 699–716.

Gitelson, R. J., & Crompton, J. L. (1983). The planning horizons and sources of information used by pleasure vacationers. *Journal of Travel Research, 21*(3), 2–7.

González Reverté, F. (2010). La celebración de eventos en Cataluña y su uso turístico. *Anales de Geografía de La Universidad Complutense, 30*(2), 107–131. Retrieved from http://revistas.ucm.es/index.php/AGUC/article/view/AGUC1010220107A/30713

Gutiérrez, J., García-Palomares, J. C., Romanillos, G., & Salas-Olmedo, M. H. (2017). The eruption of Airbnb in tourist cities: Comparing spatial patterns of hotels and peer-to-peer accommodation in Barcelona. *Tourism Management, 62*, 278–291.

Herrero, L., Sanz, J., Bedate, A., & Barrio, M. (2012). Who pays more for a cultural festival, tourists or locals? A certainty analysis of a contingent valuation application. *International Journal of Tourism Research, 14*(5), 495–512.

Kay, P. (2004). Cross-cultural research issues in developing international tourist markets for cultural events. *Event Management, 8*(4), 191–202.

Kim, K., Uysal, M., & Chen, J. S. (2001). Festival visitor motivation from the organizers' points of view. *Event Management, 7*(2), 127–134.

Li, X., Pan, B., Zhang, L., & Smith, W. W. (2009). The effect of online information search on image development. *Journal of Travel Research, 48*(1), 45–57. doi:10.1177/0047287508328659

Light, D. (1996). Characteristics of the audience for 'events' at a heritage site. *Tourism Management, 17*(3), 183–190. doi:10.1016/0261-5177(96)00005-2

Llodrà-Riera, I., Martínez-Ruiz, M. P., Jiménez-Zarco, A. I., & Izquierdo-Yusta, A. (2015). A multidimensional analysis of the information sources construct and its relevance for destination image formation. *Tourism Management, 48*, 319–328. doi:10.1016/j.tourman.2014.11.012

Mantei, C. (2012). *Concevoir la lumière comme un levier de développement touristique.* Paris: Editions Atout France.

McDowall, S. (2010). A comparison between Thai residents and non-residents in their motivations, performance evaluations, and overall satisfaction with a domestic festival. *Journal of Vacation Marketing, 16*(3), 217–233.

McHone, W. W., & Rungeling, B. (1999). Special cultural events: Do they attract leisure tourists? *International Journal of Hospitality Management, 18*(2), 215–219. doi:10.1016/S0278-4319(99)00015-8

Nicholson, R., & Pearce, D. G. (2000). Who goes to events: A comparative analysis of the profile characteristics of visitors to four South Island events in New Zealand. *Journal of Vacation Marketing, 6*(3), 236–253. doi:10.1177/135676670000600304

Quinn, B. (2005). Arts festivals and the city. *Urban Studies, 42*(5/6), 927–943.

Quinn, B. (2010). Arts festivals, urban tourism and cultural policy. *Journal of Policy Research in Tourism, Leisure and Events, 2*(3), 264–279.

Richards, G., & Wilson, J. (2004). The impact of cultural events on city image: Rotterdam, Cultural Capital of Europe 2001. *Urban Studies, 41*(10), 1931–1951. doi:10.1080/10295390208718729

Scott, D. (1996). A comparison of visitors' motivations to attend three urban festivals. *Festival Management & Event Tourism, 3*(3), 121–128.

Shanka, T., & Taylor, R. (2004). A correspondence analysis of sources of information used by festival visitors. *Tourism Analysis, 9*(1–2), 55–62.

Snelgrove, R., Taks, M., Chalip, L., & Green, B. (2008). How visitors and locals at a sport event differ in motives and identity. *Journal of Sport & Tourism, 13*(3), 165–180.

Stevens, Q., & Shin, H. (2014). Urban festivals and local social space. *Planning Practice and Research, 29*(1), 1–20.

Uysal, M., Gahan, L., & Martin, B. (1993). An examination of event motivations: A case study. *Festival Management & Event Tourism, 1*(1), 5–10.

Walsh-Heron, J., & Stevens, T. (1990). *The management of visitor attractions and events.* New Jersey: Prentice Hall.

Woodhead, C. (2017, February 19). White Night Melbourne 2017: Twelve hours of serious traipsing amid an ocean of people. *The Sydney Morning Herald.* Retrieved from https://www.smh.com.au

Fairy tale tourism: the architectural projection mapping of magically real and irreal festival lightscapes

Jane Lovell and Howard Griffin

ABSTRACT
This paper explores how established light festivals such as the Fête des Lumières in Lyon and Lumiere in Durham were first conceived by Robert-Houdin as illusory illuminations in the Loire in the 1950s. The research investigates the concept of spectacles as inversions of reality; re-situating light works within authenticity theory by exploring their manipulation of magical reality and irreality. The research uses the authors' experience of event design to assess different interactions of light with the tri-dimensional architectural canvas, suggesting three classifications of animated projection mapping events: architecturally passive, architecturally physically active and architecturally metaphysically active. Each category has implications for how spectators perceive these installations. Architecturally passive events may use fairy tale content, evoking atavistic and affective responses, the 'skinning' of buildings with magical reality is designed to evoke perceptual duality, and the wobbling unfolding of irreality may ultimately create a state of 'illuminated flow.'

Introduction

In recent years, light festivals, such as the *Fête des Lumières* in Lyon and *Lumiere* in Durham have become established events, attracting large audiences. Literature about light festivals has similarly expanded to assess their form (Edensor & Sumartojo, 2018; Lovell & Bull, 2017; Lovell, 2018; Torre, 2015) their hyperreality as simulacra (Lovell, 2018) and their consumption and atmosphere (Edensor, 2010; 2012; 2015; Edensor & Millington, 2014). This paper makes an original contribution to existing research by considering the undervalued contribution of Paul Robert-Houdin, the originator of the *son et lumière,* to the development of architectural projection mapping. The paper explores Debord's (1967 & 2014, p. 3) assertion that spectacles 'invert the real,' by focusing on links between projection mapping, magical reality and irreality, repositioning architectural light works in the context of authenticity. With the recent proliferation of light shows has come variation in the many different types of installations, all of which could be described as 'mapped- projections,' each interacting with the three-dimensional architectural canvas in varying degrees. To clarify this point, and support Edensor and Sumartojo's argument

(2018, p. 3) that light shows are not a 'homogenous mass,' this paper divides projection installations into categories termed architecturally passive, architecturally physically active and architecturally metaphysically active. Whilst there are installations that blur the lines between these categories, or perhaps occupy all three, these classifications allow a better understanding of these projections and the implications for their consumption. Research examining the intent of event creators has been said to be overlooked (Edensor & Sumartojo, 2018) and both authors have staged light works, which informs this work, providing insights into the design intentions.

Robert-Houdin and son et Lumière

Son et Lumière is the term given to the process of the dramatic lighting of buildings and external spaces, synchronised with sound and music, specifically designed to entertain. The notion of the spectacular is derived from the Baroque use of the technology of fireworks, 'their displays mirroring aspects of war explosions in their audio-visual impact' referencing military power and national identity (Lovell, 2018, p. 187). However, although the use of light to entertain predates the twentieth century (Watson, 2016), the concept of synergising light, sound and architecture was conceived much more recently. Architect Bernard Eugene (Paul) Robert-Houdin, grandson of the renowned illusionist and magician, Jean Eugène Robert-Houdin, is celebrated as the creator of the first Son et Lumière event at the Chateau de Chambord, France in 1952 (Gascar, Howard, & Robert-Houdin, 1964).

In his book, *Le Féerie Nocturne des Chateaux de la Loire* (1958), Robert-Houdin acknowledges the way that powerful searchlights used to illuminate Strasbourg Cathedral by the liberating French army in 1918, had had a strong influence on him. Although electric light had been integrated into architecture for both functional and aesthetic reasons for over forty years at that time, this symbolic illumination of Strasbourg Cathedral differed from the then typical relationship between light bulb and building. Until the early twentieth century, many buildings and structures had been adorned with multitudes of light bulbs, usually tracing outlines, features or geometries (McQuire, 2008, p. 119). Robert-Houdin, in highlighting the Cathedral as a beacon of victory for the French army ensured that the building, and not the light itself, remained the focal point, with viewers regarding the *effect* of the light, rather than the *source*. Robert-Houdin recognised the symbiotic relationship of light and architecture and as Curator of the Chateau de Chambord, he continued to use light in this way, nurturing a rapport between the audience of the 1950s and the monument of the 1550s. The shows initially consisted of multi-coloured lights directed at the chateau façade, highlighting features, whilst accompanied by live musical performances and occasionally involved theatrical performances. As he stated (1954, p. 16): 'We can do better and more at Chambord and make real fairy spectacles, playing with lumière so that we can … give ourselves to all the fantasies allowed by modern technology.' The inaugural son et lumière show was an immediate success and other displays soon followed, with Robert-Houdin exporting his expertise to create similar events across the rest of France, into Europe and the United States.

Since Robert-Houdin's initial experiments in the Loire Valley, many son et lumière events have failed to advance their ambition or scope. Indeed, the development of digital projection equipment, and the advent of mapping software in the early twenty-first century has facilitated bigger and brighter projections and enabled a proliferation of light festivals, to the extent that the sensation of 'projection mapping fatigue' (Chai, 2018) has been mooted. Cities such as Sydney, Geneva, Singapore, Stuttgart and Durham, now stage annual festivals. The participative and atmospheric qualities of these light festivals are emphasised by Edensor and Sumartojo (2018, p. 1) to dispel criticism (Mercer and Mayfield, 2015) that events such as Sydney's *White Nights* are a 'neoliberal form of entertainment,' imposed on communities as 'passively-consumed spectacles.' Torre (2015, p. 211) notes the large crowds and communal nature of projection mapping, stating that the technique promotes, 'a visibly and equally shared experience' and sensations of communitas (Turner and Turner, 1978) indicate a potentially powerful crowd atmosphere, highlighting the difference between this form of virtuality and the solitary experience of most virtual realities.

The magically real and the irreal

When the word *spectacle* is applied pejoratively to light installations, as described above, it implies broad-spectrum inauthenticity. Yet, in *The Society of the Spectacle*, Debord (1967 & 2014, p. 3) traces how 'reality emerges within the spectacle and the spectacle is real,' also noting that 'reality is inverted.' Lovell and Bull (2017, p. 8) have explored the plasticity of authenticity by applying Fjellman's (1992, p. 255) heuristic terms 'real real, fake real, real fake and fake fake' to tourists' consumption of staged and unstaged authenticity in English historic cities, (see Table 1 and Table 2). While assessing notions of fake and real and recognising that realit(ies) are culturally mediated, embodied and subjective, Lovell and Bull (2017) introduced the concept of spatial magical reality, assessing how unexpected encounters with public art made some settings seem surreal and fantastical to observers. The term magical reality is more appropriate than Robert-Houdin's magical illusions when exploring how some light installations combine the built environment and the marvellous to embody aspects of the imagination in the world. Magic realism combines '…

Table 1. Real and Fake Experiences (Lovell & Bull, 2017, p. 8).

States	*Sensations*
Real real	Heating, existential, elevated, embodied, contemplative, inhabiting the past, sublime, performative, psychoheritage, singular.
Fake real	Cooling, mechanistic, factual, distinction, status.
Real fake	Warming, enjoyment of performance, nostalgia, pretence, re-enactment, irony.
Fake fake	Freezing, critical, alienated, disappointed.
Magically real	Evaporating, enchanted, transported, poetic, the otherworldliness of the everyday
Unreal	Superheating, overwhelmed by reality which seems magnified, sublime.
Hyperreal	Melting, wobble when reality seems to be a copy, simultaneity.

Table 2. Real and Fake qualities of places (Lovell & Bull, 2017, p. 8).

States	*Qualities*
Real real	Entropic, original, ruined, gritty, slow, singular.
Fake real	Historically accurate reproduction, first-order simulacra.
Real fake	Obvious copies, pastiche, postmodern references, re-enactments, recreations of historic or fictional places, and second-order simulacra, fast, themed, decontextualised.
Fake fake	Poor copies of copies, commodified, placeless, post-truth.
Magically real	Third order simulacra, dreamlike, imaginative, artistic, rescaled, phantasmagorical, uncanny.
Hyperreal	Overwhelmingly copied.
Unreal	Singularity, iconic.
Virtually real	Circulated, viral, holographic, immersive

realism and the fantastic, so that the marvellous seems to grow organically within the ordinary, blurring the distinctions between them' Faris (2004, p. 1). The expression magic realism was first used to describe a Latin American literary genre and epitomises the disconnection between and meeting of old and new worlds in the post-colonial era (Strecher, 1999, pp. 267–269). As Arthur C Clarke (1962, p. 14) stated: 'Any sufficiently advanced technology is indistinguishable from magic,' and projection technology still seems as if it brings a wondrous future world to bear on the contemporary and past.

Magic realist author Carlos Fuentes perceived the genre architecturally in the: ' … vernacular Baroque churches, with their very austere and abstract exterior severely contrasted with the dark space of the interior with its golden, luminous retablo, which produce the quality of a dream-like experience' (Burian, 1997, pp. 50–51). According to architect Adam Sharr (2008, p. 6) 'numerous architects behave as magical realists.' He attributes to them 'strange and fantastic events which defy belief,' 'curious happenings,' and 'subversions' of practice; architectural magical realism is said to take form in the optical illusions and amorphous work of Zaha Hadid. Robert-Houdin chose the title *La Féerie Nocturne des Châteaux de la Loire* for his photographic book of light works and for him, the son et lumière clearly fell into the realm of the fairy tale as he details the fantasy of 'magic castles and enchanted palaces' (1954, p. 18). While reality is, as stated earlier, visually and culturally mediated, animated projection mapping can only further the possibilities of subversion, inversion and disruption.

In addition to magical reality, this paper introduces the category of irreality. The art curator and philosopher Goodman (1976) first used irreality to describe fantasy world-building and it has since been employed by philosophers such as Sartre (2004, p. 3), who asserted that the imagination is 'the great "irrealising" function of consciousness.' Swinford (2001, pp. 77–78) argues that irrealism's characteristics include 'mutation' and the 'usurpation of the natural.' If magical reality is the embodiment of the imagination in the everyday world, then irreality transforms the everyday in an imagined world. The term was used by Torre (2015, p. 204) to describe projection-mapped animated light installations, emphasising the impression they give of structural metamorphosis. This paper contests that some of these events are irreal and others are magically real.

Architecturally passive

Light installations interact with architecture in different ways; some incorporating elements of magical reality and others attaining full irreality (see Figure 1). Some 'mapped' projections could be said to be architecturally passive, using buildings as screens, like open cinemas. Passive light installations clearly link to their origin in the magic lantern and popularity of phantasmagoria (Benjamin, 1935 & 2002, p. 526), but without reference to the architectural space itself. At the 2017 *Fete des Lumières* in Lyon, Nathanaelle Picot and Gaël Picquet, collectively known as Pixel 'n' Pepper, displayed *Enoha fait son Cinéma* at the Place des Terreaux, using both the Hotel de Ville and the Musée des Beaux Arts as the canvas for their work (see Figure 2). The piece referenced the historic links that Lyon has with the birth and development of film, highlighting seminal cinematic moments, starting with a tribute to Georges Méliès' *A trip to the Moon,* and concluding with an allusion to Scott Derrickson's *Dr Strange*, referencing works

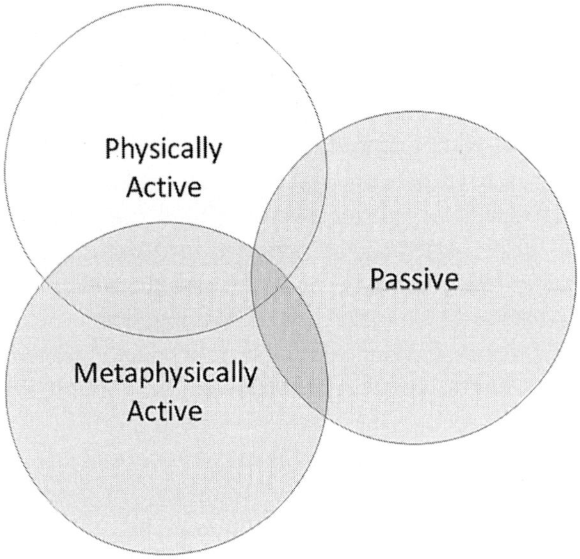

Figure 1. Projection Mapping Categories.

including Kubrick, Lucas and Spielberg. The piece predominantly used the two historic buildings as large projection screens, fleetingly keying into the architectural detail. Despite this partial interplay between the projection and the architecture, *Enoha fait son Cinéma* must be considered passive overall, a tribute to cinema, but not to the architectural components. Categorising this type of projected work as passive does not imply that there is little or no connection between the light and the surface it falls upon, but arguably, these passive works do not promote or develop the relationship between audience and building, and in fact, they tend to disregard the architectural canvas they are using.

Figure 2. Enoha fait son Cinema by Pixel 'n' Pepper.

The building acts as an illuminated manuscript, but while the designs adorn the text of the architecture, they are related to but separate on the page.

Some architecturally passive projections have a deeper relationship with the built canvas. In 2014, Gilbert Coudène's *Terre aux Lumières,* shown at the same site at the *Fête des Lumières,* extensively referenced the art contained within the Musée des Beaux Arts. The relationship fostered between animation and building in this instance, is one based on cultural reference and not on a synergy with the architectural geometry. In 2015, the Musée des Beaux Arts in Lyon attracted just over 330,000 visitors. Whereas over just four days in December 2014, an estimated 900,000 spectators stood in the Place des Terreaux to watch Coudène's work. Although few can argue against the merits of viewing the artwork in its actuality, it should be acknowledged that this projection of the art reached a far greater audience. McQuire (2008, p. 149) states: 'A crucial role for new media art in public space is the potential to avoid the filter of sites such as the art gallery, and thereby engage audiences who might never cross that threshold.' This seems to be evidenced by the attraction of between 2 and 3 million visitors to Lyon for the *Fete des Lumières,* over just a few nights each year. In addition to this, it could be argued that visitors to the museum, whilst appreciating the setting, are not there for the building itself, but for the art contained within. Conversely, those regarding the projection-mapped façade are, in a sense, there to appreciate the building, albeit in a different form to its intended appearance. In Coudene's work, the relationship between the audience, culture and architecture is deepened using projection mapping.

Civic buildings or cathedrals do not lose their sense of presence or become entirely defamiliar when experienced as projection canvases; Torre (2015, pp. 202–203) maintains that site specificity is a key ingredient in projection mapping, distinguishing it from the duplicative nature of the traditional cinematic experience. He argues that 'A … movie can play simultaneously on 3000 cinema screens anywhere in the world, and thanks to the generic nature of the screen (being smooth, flat and white) the movie will look more or less comparable regardless of where it is shown.' When describing the use of monumental heritage structures for light projection, Lovell (2018, p. 191) refers to their locational narrative and 'immanent and imposing presence.' In some cases, the building 'maps the designer' with its architectural foibles and significance, thereby affecting the design possibilities (Lovell, 2018, p. 189).

Although the installation could be described as architecturally passive, the experience may still be affective; audiences co-create and assess experiences. If architecturally passive projections may be considered cinematic, McGowan (2012, p. 74) describes how film audiences are 'open to the gaze,' projecting their emotions onto the screen, although the object of desire is not physically there. Robert-Houdin (1954, p. 22) wrote of the 'enchantment of night-time fairy stories' and fairy tales draw on deeper-seated atavistic fears which can be summoned with the silhouettes and shadows of magic lanterns. The 'technological uncanny' Collins and Jervis (2008, p. 1) caused by the shadows created by urban lighting and the timing of events at the liminal, gloaming 'creepy time,' (Edensor & Sumartojo, 2018, p. 12) contribute to audience suggestibility. Some lighting companies and artists consciously *design in* magical reality, using fantasy content, which may be unrelated to the architecture. As the company LCI Productions delineates, discussions with local stakeholders during the planning of an event at Conwy Castle encouraged them to weave local distinctiveness into the content in a fairy tale format in acts of myth-making. During the show, a unicorn and mushrooms sprout, a crow makes princesses vanish, gingerbread men wave, and a giant silhouette of a woodcutter with an outsized scythe slices through a Sleeping Beauty forest in a form of 'Disney reality' (Fjellman, 1992, p. 60). As Lovell (2018) and Torre (2015) have observed, the renowned light show at Santiago de Compostela similarly draws on mythology, featuring dragons and forests reclaiming the building in a post-anthropocentric world. In addition to magical content, the effects may bend reality in a Harry Potteresque twist, for example, in their work *Aquarium*, at London Lumiere (originally shown at Durham Lumiere in 2014). Factory artists Benedetto Bufalino and Benoit Deseille turn an iconic red telephone box into an aquarium of exotic fish. These installations are consciously designed as magically real, myth-making what could be termed the 'heritage marvellous.'

Architecturally physically active

Architecturally physically active works take technique and effect a step further, truly disrupting the night. In February 2018, projection artist Ross Ashton and sound designer Karen Monid displayed their installation *Line* at the Cheriton Light Festival in England. The work, involving the projection of light and dark shapes onto the surface of a small church, sought to create an 'architectural metamorphosis.' The installation had previously been shown at Cambridge and Napa, California, albeit mapped to different geometries. They provide a luminous cloaking of the building, which neither seeks to deceive nor conjure illusion, but merely aims to clothe the built form in a new skin. The skin developed by physically active installations resembles 'Second Space,' or 'un espace proper' (Lefebvre, 1991), a pure space of intellect, projecting the vision of the projection-mapping architect and the original architect. Patrice Warrener, a regular contributor to light festivals over many years, also uses light in this way, creating polychromatic overlays that alter the way we look at a building or façade. *Chromolithe*, the name Warrener gives to this process, exemplifies this notion of re-skinning a building. Projection mapping, has an opportunity to harness a symbiotic relationship between the image and the building; 'concretizing the animation', as Torre (2015, p. 205) argues, giving depth to two-dimensional images. While the 'skin of light' covers the building, the projections entwine with it like tattoos, saturating blank surfaces with colour. In an event, *Cathedrals of Light*, with which one of the authors was involved, Monet's series of paintings of Rouen Cathedral were aligned to and projected back onto the exterior of the building. The sum of paint, stone and light was physically active, unlike Coudene's Musée des Beaux Arts projection. Instead, Monet's intangible 'enveloppe' of the same light on the same building at different times of day (Thomson, 2018, p. 165) was revealed, melting the cathedral by abstracting it; recreating not the painting, but the essential *en plein air* illusions of impressionism.

Far from being virtual, the 'enveloppe' effect captured in the paintings is mirrored in the way that the projected augmentation of buildings fosters a physical connection through skinning. The interrelationship between the projected photons of light and the material surface of the architecture creates a bond not experienced through other means of virtual augmented reality. It is essential to note that this connectivity between light and architecture is physical, the photons either being reflected off, or absorbed into the material surface of the building. This microscopic infusion represents the actuality of projection mapping, setting it aside from other forms of augmented reality, which simply place digital filters between the eyes of the viewer and the recorded space in the image. Pioneers of projection mapping originally used the term Spatial Augmented Reality, and this phrase perhaps better describes the interplay.

In September 2017, the technique of skinning was appropriated for the South-West Porch entrance of Canterbury Cathedral in a project organised by one of the authors (see Figure 3). The experimental work aimed to test the way in which precise masking of projected video textures could highlight the elaborate nature of the architectural detail, and thereby draw attention to them. This masking allowed elements of the façade, such as the canopy, the framework and the statues to be highlighted, which when illuminated, appeared to glow skeletally in the dark. It could be argued that this isolating of each element assists the viewer to appreciate the detail, and, unlike the experience during daylight hours, avoids the distraction of the whole and makes the material magical. Myopic architectural appreciation of individual features forms a key part of the experience.

As this paper has discussed, the term spectacle has been used sweepingly by critics to describe light festivals. Lovell (2018) has suggested that projection-mapped installations may also be defined as simulacra (Baudrillard, 1981), moving from first to third-order as they are performed. In fact, different events may resemble different types of simulacra; for example, the process of skinning could be said to be a hyperreal (Baudrillard, 1981; Eco, 1986) technique, creating a sensory and physical light copy of the building, suggestive of a second-order simulacrum, or fantastical, phantasmagorical copy which supersedes the original structure (for the duration of the event). Leal (1995, p. 101) states that: 'In magical realism the writer confronts reality and tries to untangle it' and lighting designers turn this concept on its head; they tangle realities using skinning. Rather than passively spectating, the effect on the viewer is potentially active and hyperreal; skinning is designed to overwhelm the senses, inverting and wobbling reality. Instead of being 'dazzled' by electronics into machine-like states, as Diack (2012, p. 11) suggested, when the inanimate is animated, viewers may switch on the 'illuminated gaze' (Lovell, 2018; adapted from Urry, 2002), which is open to enchantment and the marvellous, leading to vertiginous sensations of wonder and immersion. Strecher (1999, pp. 267–269) observes that magic realism is 'a worldview that permits the "magical" to coexist with the "real",' and both light and architecture have equal stakes; the illuminated gaze retains the duality of magical reality, the building remaining beneath the effect, made marvellous.

Figure 3. Canterbury Cathedral Southwest porch with illuminations by Griffin.

Architecturally metaphysically active

Active projections have the capacity to interact with architecture on a metaphysical level, as well as physical, irrealising, de-situating and erasing the historicity of buildings. These projects could be said to abstract, reinterpret and resituate structures in the realm of the lived imagination. The concept of liminality (Turner, 1973) captures the movement from one stage to another and architecturally metaphysical installations move from real to irreal, creating liminal zones. In 2012, Canterbury Christ Church University hosted Golden Night, part of their 50th Anniversary celebrations. Lianne Clark, a student at the University of Kent, produced a series of projections that sought to interact with an unremarkable administrative campus building. Devoid of architectural decoration, detail and, some would argue, character, the Fisher Building, provided a simple, yet geometrically strong backdrop for Lianne's work. Her series of animations interplayed with the geometry of the building, alluding to moving windows, bouncing façades and spinning floors. Torre (2015, p. 211) refers to the Deleuzian concept of the fold, citing how the building has a relationship with itself, for example, 'the interior of a form is merely the folding of its exterior.' The unfolding of the irreal building, seems to reveal, as Merleau-Ponty (1961, p. 132) details, the 'inside of the outside and outside of the inside.' With over 3,000 people attending the event to witness this architectural transformation, this example of metaphysical projection mapping demonstrates the way it can draw attention to even the most benign of buildings in a compelling way.

In 2016, at the *Fete des Lumières*, Yann Nguema and EZ3kiel worked together to create *Evolutions* (see Figure 4), an installation incorporated into the façade of the Cathédrale de St. Jean. The award-winning piece is fully realised Spatially Augmented Reality. The Cathedral is a well-used canvas during the festivals, but it could be argued that Nguema's work takes the notion of active projection further than the typical work seen

Figure 4. Evolutions by Nguema and Ez3kial.

each December. *Evolutions* was a poetic interpretation of the history of the cathedral, alluding to a wider conflict between science and religion. The intricate gothic architectural detail appears to warp and distort, as if influenced by immense external and internal forces, causing the façade to ripple and deform. The work continually interacts with the façade, creating a fully realised metamorphosis of the architecture which crosses over from solidity into animation (Torre, 2015, p. 204). This technique is used by other designers such as AntiVJ and could be argued to be the zenith of projection mapping, where both building and light explode into a form which is neither solely architecture, nor solely light. It is no longer a building made marvellous, it is a lived imaginary and third-order simulacrum, which *precedes* the original, building on Paul Robert-Houdin's intentions by making the imaginary metamorphosing building, rather than the light, the focal point.

Metaphysically active installations are the trompe l'oeil of the modern day and like painted ceilings they are aimed to deceive with tricks of perspective. The illusory effect that forms that basis of this active projection mapping stems from a manipulation of the architectural geometry, tricking our perception of form and space. With the X and Y (read width and height) layout of an architectural façade fixed in place (see Figure 5), much of the illusion of movement typically stems from a perceived manipulation of the Z-axis. This apparent movement in the third dimension is usually produced by the creation of imaginary form or the implied subtraction of the architecture to create space. As Robert-Houdin (1954, p. 9) writes of the way in which he tries to use light: '... it is enough to have only one arm of a Greek statue to imagine its perfection, it only needs a keep or a wall to evoke the disappeared castle.' This extrusion out from and intrusion

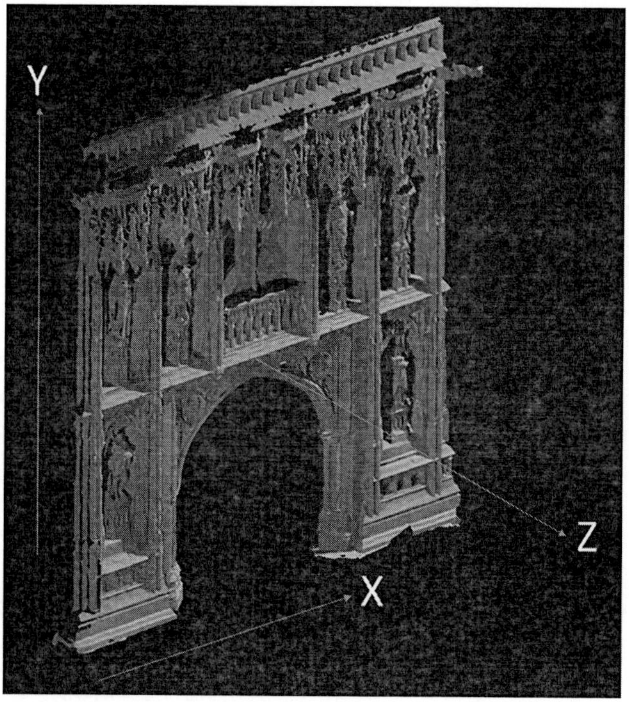

Figure 5. 3D scan of Canterbury Cathedral SW porch showing *X*, *Y* and *Z* axes.

into a façade, constitutes a core feature for this type of deceptive work taking the work beyond architectural appreciation. In *Evolutions*, Nquema uses this notion of extrusion to produce an incredibly persuasive illusion (see Figure 6). However, the apparent extrusion from and intrusion into architectural facades, what could be described as a convincing application of a fake third dimension onto a façade is dependent upon the vantage point of the spectator or spectators. The further the stretching of the Z-Axis that occurs, the more critical it is to be viewed from the correct position, commonly known as the 'sweet-spot.' In February, at Cheriton Light festival, the author explored the issue by creating imagery designed to be viewed from a specific vantage point. *A to Z-Axis* used both intrusion and extrusion to imply a fake third dimension, specifically aimed at spectators in a neighbouring street, so that they could be drawn into this deception. Figure 7 shows the skewed image projected and the resulting view from the sweet-spot.

The irreality of architecturally metaphysically active projection has implications for the light event experience, by (dis) embodying spatial creativity and fluidity, on a non-human scale. Rescher (2003, pp. 165–168) proposes that: 'Our cognitive access to irreality is limited to the suppositional projection of possibilities and these never manage to identify authentic individuals or worlds.' He goes on to say that: 'Nonexistents ... do not have a claim on actuality of some sort. They are, by definition, something unreal; they are mere objects.' However, light, animation, software, hardware, could be argued to be part of connected experiences. Drawing on actor-network theory, Lugosi and Quinton (2018) argue that technology and the products and illusions associated with it are part of the more-than-human world, which could be equated to what Rescher (2003) describes as 'irreal materiality.' The notion of irreality was discussed by Sartre in relation to the imaginative consumption of artworks, which share commonalities with projection-mapped animation. De Warren (2013, p. 99) qualified Sartre's conception of the irreal when

Figure 6. Evolutions by Nguema and Ez3kial.

Figure 7. A to *Z*-Axis by Griffin.

regarding paintings as a state emerging from the combination of the subject of the painting, the painted object and the act of imagination when viewing the painting. As Sartre states: (2004, p. 190) 'the painting should be conceived as a material thing visited from time to time (every time that the spectator takes the imagining attitude) by an irreality that is precisely the painted object.' Thus, when we look at a painting, to evoke the 'presence' of the subject of the painting, the painting ceases to be material and is absented, to be replaced by irreality. In *The Two versions of the Imaginary*, Blanchot (1981) also refers to the manner in which the image follows the object; the object must be displaced to 'elsewhere' in order to grasp the image in our imagination. The metaphysically active light show is a third version of the imaginary, the building sublime, as we have never dared to imagine, radiating impossibilities beyond the mere marvellous to spin windows, liquefy and warp stone with superirrealism.

While the audience is aware that the building does not strictly have agency, irreal materiality gives buildings what De Warren (2013) argues is 'the spell of a presence, as if the medium surrendered its materiality to the spectral body of the imaginary.' The 'spell of presence' (a preoccupation of Robert-Houdin's grandfather, Jean Eugène, who gave machines the illusion of life as automatons) is created by displacing the physical building with representation, conveying an impossible structure with its own 'imagining attitude.' Smith (2015, p. 59) discusses the *Animation Culturel* events in France in the 1970s as an example of how events became central to urban cultural strategies, referring to the way in which the word animation was used in the sense of 'giving life to', or 'to inspirit' public space. The 'presence' of irreal materiality is linked to the sentience of place, discussed by Lovell and Bull (2017, p. 13) drawing on Schwenger's (2006, pp. 37–40) assessment of the diffuse Lacanian sensation of objects 'looking back.' Architecturally metaphysically active displays emphasise the intersubjective gaze, enabling the viewer to, as Merleau-

Table 3. Passive, active and metaphysical architectural projection mapping.

Architectural typology	Characteristics	Consumption	State of reality
Passive	Architecture acts as canvas, building and light are separate, the light dominating.	Spectating, gazing at the light effects.	Magically real
Physically Active	Optical illusions, architectural skinning of the building with light, causing duality; the light and building are equal, superseding reality.	Duality of perceptual shifts, illuminated gaze.	Magically real
Metaphysically Active	The building becomes animated and has agency, the metamorphosing form *preceding* reality, dominating both light and building, which are absent, replaced by a parallel world.	Illuminated flow, lived experience of material irreality.	Irreal

Ponty (1961, p. 132) observes, 'see with it' rather than looking *at it*, or *being looked at*. As Lovell (2013, p. 190) applied Csikszentmihalyi's (1988, p. 416) concept of the 'flow' of optimal experiences, to define 'heritage flow;' a 'transportion' (Green & Brock, 2000) to a 'past place,' when tourists visiting heritage cities enter an imaginative, non-representation state. During actively metaphysical light installations, the 'illuminated gaze' may retain its characteristics of receptiveness and wonder, developed from the realm of the magically real, becoming a transformative 'illuminated flow' as the viewer is absented, deep in the irreal, 'elsewhere.'

Conclusion

This paper uses case studies of different projection-mapped events staged or attended by the authors, illustrating from this perspective how, while our individual realities may differ, light installations can invert and disrupt them further, suggesting that some installations offer qualities of experience which are far from the 'passively consumed spectacles' or 'projection mapping fatigue' dismissed by critics. The three approaches to architectural projection mapping have been clearly delineated in Table 3 and the updated typologies of reality are outlined in Table 4. Both magic realism and irreality are terms with implications for the authenticity of light festival staging and consumption, which can trace their lineage not simply technically, but conceptually from the 'magical illusion' of Robert-Houdin's *Le Féerie Nocturne des Chateaux de la Loire*.

Many passive works remain two-dimensional in nature, using buildings as screens in Baroque displays of illusory power and employing content unrelated to the architecture and its dreams. The physically active displays invert reality, evoking the marvellous

Table 4. Real and Fake Experiences and Qualities of Light Shows (Adapted from Lovell & Bull, 2017, p. 8).

States	*Sensations*
Magically real	Evaporating, enchanted, transported, the otherworldliness of the everyday, embodied imagination, marvelling, duality.
Hyperreal	Melting, wobble when reality seems to be a copy, inverted reality, simultaneity.
Irreal	Metamorphosing, world-building, fantasy, imagining attitude, irreal materialisations.
States	*Qualities*
Magically real	Dreamlike, imaginative, artistic, phantasmagorical, marvellous, fairy tales, mythical, skinning.
Hyperreal	Overwhelmingly copied, rescaled, spectacular, inverted reality.
Irreal	Third order simulacra, metamorphosis, mutation, fantasy, world-building, spell of presence, agency, irreal materiality.

through the skinning of buildings with colourfully tattooed, perceptual shifts, establishing a duality between light and the architecture. It seems to the authors that magic realism's embodiment of the imagination in the real world epitomises both passive and physically active forms of display, because in both typologies, the building flickers beneath. In contrast, architecturally metaphysically active installations are capable of intrinsically linking light and architecture to displace the static building beneath, giving the 'presence' of structural agency, three-dimensional geometrical shifts and possibilities. The research makes a further original contribution to the study of experience by indicating that the illuminated gaze becomes a more transformative state of 'illuminated flow' during irreal performances as the spectator looks *into* rather than *at* the light effects. This last category is the third-order simulacrum, *preceding* reality with material irreality through a quantum unfolding of architectural light-self, vanishing of 'elsewhere,' when no-one is there to observe it, leaving the building forever ambiguous.

Disclosure statement

No potential conflict of interest was reported by the authors.

References

Baudrillard, J. (1981). *Selected writings: Simulacra and simulation*. Cambridge: Polity Press.

Benjamin, W. (1935 & 2002). *The arcades project*. R. Tiedemann, (Ed.). (H. Eiland & K. McLaughlin, Trans.). London: The Belknap Press.

Blanchot, M. (1981). *Two versions of the imaginary. The gaze of Orpheus and other literary essays*. (L. Davis, Trans.). New York: Station Hill Press. pp. 79–89.

Burian, E. (1997). *Modernity and the architecture of Mexico*. Austin: University of Texas Press.

Chai, A. (2018, January 25) Tired of projection mapping? Well it's nowhere near its peak yet, say artists. *SG Magazine*. Retrieved from http:/sg.asia-city.com/things-do/news/tired-projection-mapping-well-its-nowhere-near-its-peak-yet-say-artists

Clarke, A. C. (1962, 2013). *Hazards of prophecy: The failure of imagination*. London: Orion Publishing Group.

Collins, J., & Jervis, J. (2008). Introduction. In J. Collins & J. Jervis (Eds.), *Uncanny modernity: Cultural theories, modern anxieties* (pp. 1–9). London: Palgrave.

Csikszentmihalyi, M. (1988). The future of flow. In M. Csikszentmihalyi & I. Csikszentmihalyi (Eds.), *Optimal experience: Psychological studies of flow in consciousness* (pp. 364–383). New York, NY: Cambridge University Press.

De Warren, N. (2013). Flesh made paint. *Journal of the British Society for Phenomenology*, *44*(1), 78–104.

Debord, G. (1967 & 2014). *The society of the spectacle*. (Knabb, K. Trans.). Berkley, CA: Bureau of Public Secrets.

Diack, H. (2012). Sleepless nights: Contemporary art and the culture of performance. *Public. 23*(45), 8–22.

Eco, U. (1986). *Faith in fakes: Travels in hyperreality*. London: Picador.

Edensor, T. (2010). Aurora landscapes: Affective atmospheres of light and dark. In K. Benediktsson & K. A. Lund (Eds.), *Conversations with landscape* (pp. 227–240). Abingdon: Routledge.

Edensor, T. (2012). Illuminated atmospheres: Anticipating and reproducing the flow of affective experience in Blackpool. *Environment and Planning D: Society and Space*, *30*(6), 1103–1122.

Edensor, T. (2015). Light design and atmosphere. *Visual Communication*, *14*(3), 331–350.

Edensor, T., & Millington, S. (2014). Blackpool. In S. Isenstadt, M. M. Petty, & D. Neumann (Eds.), *Cities of light: Two centuries of urban illumination* (pp. 58–66). Abingdon: Routledge.

Edensor, T. & Sumartojo, S. (2018). Reconfiguring familiar worlds with light projection: The Gertrude street projection festival, 2017. *GeoHumanities*, *4*(1), 112–131.
Faris, W. (2004). *Ordinary enchantments: Magic realism and the remystification of narrative*. Nashville, TN: Vanderbilt University Press.
Fjellman, S. (1992). *Vinyl leaves: Walt Disney World and America*. Boulder, CA: Westview Press.
Gascar, P., Howard, R., & Robert-Houdin, P., (1964). *Chambord*. London: Collier Macmillan.
Goodman, N. (1976). *Languages of art: An approach to the theory of Symbols*. Indianapolis, IN: Hackett.
Green, M. C., & Brock, T. C. (2000). The role of transportation in the persuasiveness of public narratives. *Journal of Personality and Social Psychology*, *79*(5), 701.
Leal, L. (1995). Magic realism in Spanish America. In Zamora, L. P. & Faris, W. (Eds.), *Magic realism: Theory, history, community*. London: Duke University Press.
Lefebvre, H. (1991). *The production of space*. Oxford: Blackwell.
Lovell, J. (2013). *Cities of imagination: Staged authenticity in historic cities* (Doctoral dissertation, University of Kent).
Lovell, J. (2018). Hyperreal light Simulacra: Performing heritage buildings. In J. Rickly & E. Vidon (Eds.), *Authenticity & tourism: Materialities, perceptions, experiences* (pp. 181–200). Bingley: Emerald.
Lovell, J. & Bull, C. (2017). *Authentic and inauthentic places in tourism: From heritage sites to Theme Parks*. Abingdon: Routledge.
Lugosi, P., & Quinton, S. (2018). More-than-human netnography. *Journal of Marketing Management*, *34*(3–4), 1–27.
McGowan, T. (2012). *The real gaze: Film theory after Lacan*. New York, NY: SUNY Press.
McQuire, S. (2008). *The media city: Media, architecture and urban space*. London: Sage.
Mercer, D., & Mayfield, P., 2015. City of the Spectacle: White Night Melbourne and the politics of public space. *Australian Geographer*, *46*(4), 507–534.
Merleau-Ponty, M. (1961). Eye and mind. In S. Manghani, A. Piper, & J. Simons (Eds.), *2006. Images: A reader* (pp. 131–134). London: Sage.
Rescher, N. (2003). *Imagining irreality: A study of unreal possibilities*. Chicago, IL: Open Court.
Robert-Houdin, P. (1954). *La Féerie Nocture des Châteaux de la Loire*. Paris: Librairie Hachette.
Sartre, J.P. (2004). *The imaginary: A phenomenological psychology of the imagination*. (Webber, J. Trans.). London: Routledge.
Schwenger, P. (2006). *The tears of things: Melancholy and physical objects*. London: University of Minnesota Press.
Sharr, A. (2008, November). *Strange decisions: Subversiveness and magical realism in architecture*. Paper delivered at the Architectural Humanities Research Association Conference: *Before and after Agency*. Sheffield.
Smith, A. (2015). *Events in the city: Using public spaces as event venues*. Abingdon: Routledge.
Strecher, M. C. (1999). Magical realism and the search for identity in the fiction of Murakami Haruki. *Journal of Japanese Studies*, *25*(2), 263–298.
Swinford, D. (2001). Defining irrealism: Scientific development and allegorical possibility. *Journal of the Fantastic in the Arts*, *12*(1), (45), 77–89.
Thomson, R. (2018). *Monet & architecture*. London: National Gallery.
Torre, D. (2015). The metamorphosis of place: Projection-mapped animation. In C. Pallant (Ed.), *Animated landscapes: History, form and function* (pp. 199–214). London: Bloomsbury Publishing USA.
Turner, V., 1973. The center out there: Pilgrim's goal. *History of Religions*, *12*(3), 191–230.
Turner, V., & Turner, E. (1978). *Image and pilgrimage in Christian culture*. New York: Columbia University Press.
Watson, B. (2016). *Light: A radiant history from creation to the quantum age*. New York, NY: Bloomsbury.
Urry, J. (2002). *The tourist gaze*. (2nd ed.). London: Sage.

Index

annual light event 86
anti-social behaviour 5, 53, 67
architectural projection mapping 99, 111
architecture 89–90, 96, 100, 102–108, 111–112; metaphysically active 107; passive 102
astro-tourism 4
authenticity 67–70, 75, 77, 79–80, 82–83, 99, 101, 111

Bissell, D. 42
Blanchot, M. 110
Bouchon, F. 56
Budapest 6, 25–30, 32–33
Bull, C. 101, 110

candlelight 73, 77, 79, 82
Chalip, L. 89
Church 79–80
city tourism 2
Clarke, Arthur C. 102
clubs 4, 39, 46, 52–55, 57, 59–62, 67
collective strategies 46–47
collegial relations 46
Collins, J. 104
Colomb, C. 5, 13
commodification 3, 5, 13, 19, 28
conflicts 5–6, 12, 27, 32–33, 53–56, 61
contemporary tourist city 3, 5, 13, 19, 28
Cresswell, T. 12, 18
cultural capital 58–61
cultural events 77–78, 85, 87–89, 96
cycling 43–44

data collection 71
Debord, G. 99
Demers, A. 86–87
De Warren, N. 109–110
Diack, H. 106
Dirksmeier, P. 57
Douglas, Mary 17

Edensor, T. 69, 77, 99
electric light 73, 100
emotions 2, 11, 38, 60, 67–69, 75, 104
entrepreneurs 27, 30, 32–33
ephemeral illumination 85–86, 88
Evans, G. 68
existential authenticity 69, 80
extrusion 108–109

fairy tale tourism 99
festival 2–3, 7, 79, 86–96, 107
festival lightscapes 99
Fjellman, S. 101
Florida, R. 54
French night events 70
Füller, H. 56

Garcia, L. M. 57
gentrification 2, 24–26, 28, 33, 54–55, 57, 62
Getz, D. 87
Giordano, E. 88
Goodman, N. 102
Green, B. 89
Gustafson, P. 62

Hadid, Zaha 102
Hannam, K. 58
Helbrecht, I. 57
Hernández, B. 60
Herrero, L. 95–96
hospitality 30, 38–40, 44, 46–47
hospitality workers 37–38, 40–41, 43–44, 47–48

illuminated gaze 106, 111–112
illumination 5, 68, 85–87, 94, 96

imagination 101–102, 110, 112
information sources 90–91, 93, 95
irreality 99, 102, 109–111
itineraries 70–72, 77, 89–91
Iveson, K. 58, 89

Jervis, J. 104
Jones, Owen 16

Kreitzman, J. 4

Laussu, C. 81
Lawler, S. 17
Leal, L. 106
leisure economy 24–26
Light, D. 88–89, 95
light-based events 5
light effects 112
light festivals 5, 7, 68–69, 73, 79, 81, 85, 87–90, 95–96, 99, 101, 105–106
light installations 101, 111
light/lighting events 86–87
light projections 5, 7, 104
Lim, S. E. Y. 56
Lloyd, R. 54
Llum BCN festival 86, 89–91, 96
London 4, 26, 53, 55, 57, 67, 88
Lourdes 7, 66, 70–71, 73–81
Lovell, J. 101, 104–106, 110–111
low-cost nocturnal tourism 10–12, 15
low-cost tourism 11–12
low-cost tourist-oriented nightlife 11
ludification 12, 19
Lugosi, P. 109
Lyon Festival of Lights 68, 70, 77–78, 80, 82

MacCannell, D. 68, 79
Madrid 6, 10–11, 13–15
magical reality 99, 102, 104, 106
Mantei, C. 86, 88
McDowall, S. 89
McGowan, T. 104
McHone, W. W. 88, 95
McQuire, S. 104
Merleau-Ponty, M. 110–111
Michel, B. 56
middle-class culture 12, 15, 17
middle-class residents 11, 16–18
mobilities 2, 12, 38–39, 41–42, 47–48, 56, 58–60, 62
modern global cities 58
monuments 85–86, 94, 100
motivations 6–7, 70, 80, 89–92, 95

neoliberalism 6, 25–27, 30, 32
neo-patrimonialism 26, 29
new urban tourism 52–53, 55–56
New York 3, 5, 26, 55, 57
Nicholson, R. 88
nightlife 2–6, 12–13, 16, 52–55, 59, 61, 67–68; consumers 53–54, 56–57; spaces 55, 59–60; tourists 57
night-time economy 6, 11, 15, 24, 37, 47, 68
night-time tourism 3
nocturnal ritual activities 66, 73
nocturnal ritual practices 67, 73, 77–78, 81–82
Nofre, J. 57
Novy, J. 5

Ong, C. E. 88

Page, S. J. 87
Pearce, D. G. 88
permanent illuminations 5, 85
pilgrimage 70–73, 80; cities 66, 80, 82
pilgrims 7, 66, 70–72, 75, 77, 82
police 14, 29–31
post-industrial cities 12, 19
post-socialist cities 24–25, 27–28
projection mapping 99, 101, 104–106, 108
protests 10–11, 13, 18–19
public transport 38, 41–42, 44, 62

Quinn, B. 88
Quinton, S. 109

regulation 6, 24, 26–27, 29–31
religious events 69–70, 79, 87
Rescher, N. 109
Richards, G. 87
Robert-Houdin, P. 100, 104, 108
Rungeling, B. 88, 95

sacred 7, 66–67, 69–70, 72–73, 75, 77–80, 82; places 66, 69–70, 73, 82
Sartre, J.P. 110
satisfaction 89–91, 94–95
Schwenger, P. 110
seasonal illumination 85–86
secular visitors 75
sense of place 78–79
Shanka, T. 95
Sharr, Adam 102
Shaw, R. 4
shrine towns 70–71, 73, 77–80, 82
silence 68, 79, 82
Smith, A. 110

Snelgrove, R. 89, 95
Spracklen, K. 15
Strecher, M. C. 106
Sumartojo, S. 99
Swinford, D. 102

Taks, M. 89
Taylor, R. 95
theatres 3–4, 67–68, 87
Thurnell-Read, T. 13
Tomaney, J. 18
torchlight processions 67–68, 70, 73–74, 77–78, 81–82
Torre, D. 102, 105, 107
tourist city 11–12
tourist experiences 6–7, 53, 67–70, 73, 79–80, 82
tourists 3–7, 10, 17–18, 29, 53, 55–58, 61–62, 68–69, 71–73, 75, 77, 79–80, 82, 87–89

urban spaces 3, 5, 14, 19, 43, 86–88
urban tourism scholars 6, 56, 61–62
urban tourists 6, 55, 57

visitor behaviour 88–89
visitor profiles 88–89
visitors 4, 6, 11, 13, 57, 66–80, 82, 85–86, 88–96, 104
visitor's consumption 13

Wales' Licensing Act 53
Wilson, J. 87
work 2–4, 6, 13, 37–38, 40–46, 48, 54, 56–57, 59, 100, 102, 105, 107–109
workers 6, 38–42, 44, 46–47, 54
workplace 38, 40, 45–46; relations 46–47

Zukin, Sharon 11